**Marcelo Nogueira de Sousa**

# Multipath Exploitation for Emitter Localization using Ray-Tracing Fingerprints and Machine Learning

# Multipath Exploitation for Emitter Localization using Ray-Tracing Fingerprints and Machine Learning

Marcelo Nogueira de Sousa

Universitätsverlag Ilmenau
2021

## Impressum

**Bibliografische Information der Deutschen Nationalbibliothek**
Die Deutsche Nationalbibliothek verzeichnet diese Publikation in der Deutschen Nationalbibliografie; detaillierte bibliografische Angaben sind im Internet über http://dnb.d-nb.de abrufbar.

Diese Arbeit hat der Fakultät für Elektrotechnik und Informationstechnik der Technischen Universität Ilmenau als Dissertation vorgelegen.

| | |
|---|---|
| Tag der Einreichung: | 15. Oktober 2020 |
| 1. Gutachter: | Univ.-Prof. Dr.-Ing. habil. Reiner S. Thomä<br>(Technische Universität Ilmenau) |
| 2. Gutachter: | Univ.-Prof. Dr.-Ing. Klaus Witrisal<br>(Graz University of Technology, Österreich) |
| 3. Gutachter/-in: | Prov. Dr. José Antônio Apolinário Junior<br>(Military Institute of Engineering, Rio de Janeiro, Brazil) |
| Tag der Verteidigung: | 16. April 2021 |

Technische Universität Ilmenau/Universitätsbibliothek
**Universitätsverlag Ilmenau**
Postfach 10 05 65
98684 Ilmenau
http://www.tu-ilmenau.de/universitaetsverlag

**ISBN** 978-3-86360-244-4
**DOI** 10.22032/dbt.49241
**URN** urn:nbn:de:gbv:ilm1-2021000143

---

Titelfoto: photocase.com | AlexFlint

# Abstract

The precise localization of radio frequency (RF) transmitters in outdoor environments has been an important research topic in various fields for several years. Nowadays, the functionalities of many electronic devices are based on the position data of a radio-frequency transmitter using a *Wireless Sensor Network* (WSN). Spatially separated sensors can measure the signal from the transmitter and estimate its location using parameters such as *Time Of Arrival* (ToA), *Time Difference Of Arrival* (TDOA), *Received Signal Strength* (RSS) or *Direction Of Arrival* (DOA). However, certain obstacles in the environment can cause reflection, diffraction, or scattering of the signal. This so-called *multipath effect* affects the measurements for the precise location of the transmitter. Previous studies have discarded multipath information and have not considered it valuable for locating the transmitter. Some studies used ray tracing (RT) to create position fingerprints, without reference measurements, in a simulated scenario. Others tested this concept with real measurement data, but this proved to be a more cumbersome method due to practical problems in the outdoor environment.

This thesis exploits the concept of *Channel Impulse Response* (CIR) to address the problem of precision in outdoor localization environments affected by multipath. The study aims to fill the research gap by combining multipath information from simulation with real measurements in a machine learning framework. The research question was whether the localization could be improved by combining real measurements with simulations. We propose a method that uses the multipath fingerprint information from RT simulation with reference transmitters to improve the location estimation. To validate the effectiveness of the proposed method, we implemented a TDoA location system enhanced with multipath fingerprints in an outdoor scenario. This thesis investigated suburban and rural areas using well-defined reflective components to characterize the localization multipath pattern. The results confirm the possibility of using multipath effects with real measurements to enhance the localization in outdoor situations. Instead of rejecting the multipath information, we can use them as an additional source of information.

# Zusammenfassung

Die präzise Lokalisierung von Hochfrequenz (HF)-Sendern im Außenbereich ist seit einigen Jahren ein wichtiges Forschungsthema in den verschiedensten Bereichen. Heutzutage basieren die Funktionsweisen vieler elektronischer Geräte auf den Positionsdaten eines HF-Senders unter Einsatz eines *drahtlosen Sensornetzwerks* (Wireless Sensor Network). Räumlich getrennte Sensoren können das Signal des Senders messen und seinen Standort mit Parametern wie *Ankunftszeit* (Time Of Arrival), *Ankunftszeitdifferenz* (Time Difference Of Arrival, TDoA), *empfangene Signalstärke* (Received Signal Strength) oder *Ankunftsrichtung* (Direction Of Arrival) abschätzen. Jedoch können bestimmte Hindernisse in der Umgebung Reflexion, Beugung oder Streuung des Signals veursachen. Dieser sogenannte *Mehrwegeffekt* beeinträchtigt die Messungen für die präzise Lokalisierung des Senders. Frühere Studien haben die Mehrweginformationen verworfen und sie nicht als wertvoll für die Lokalisierung des Senders angesehen. Einige Studien verwendeten *Ray-Tracing* (RT), um in einem simulierten Szenario Positions-Fingerprints, ohne Bezug zu realen Messungen, zu erstellen. Andere wiederum testeten dieses Konzept mit realen Messdaten, was sich jedoch aufgrund der praktischen Probleme im Außenbereich als umständlicheres Verfahren herausstellte.

Diese Doktorarbeit befasst sich mit dem Problem der präzisen Lokalisierung von HF-Sendern in Außenbereichen, welche von Mehrwegeffekten betroffen ist, unter Verwendung der *Kanalimpulsantwort* (Channel Impulse Response). Die Arbeit zielt darauf ab, die Forschungslücke zu schließen, indem die Mehrweginformationen aus der Simulation mit realen Messungen in einem maschinellen Lernrahmen kombiniert werden. Die Fragestellung war, ob die Lokalisierung durch die Verknüpfung realer Messungen mit Simulationen verbessert werden kann. Wir schlagen eine Methode vor, die die Mehrweg-Fingerprint-Informationen aus der RT-Simulation mit Referenzsendern nutzt, sodass die Standortbestimmung verbessert werden kann. Um die Wirksamkeit der vorgeschlagenen Methode zu validieren, haben wir ein TDoA Lokalisierungssystem in einem Außenbereich-Szenario implementiert. Im Rahmen dieser Dissertation untersuchten wir vorstädtische und ländliche Gebiete unter Verwendung von wohldefinierten reflektierenden Komponenten zur Charakterisierung des Mehrwegmusters. Die Ergebnisse bestätigen die Möglichkeit, Mehrwegeffekte mit realen Messdaten zu nutzen, um die Lokalisierung im Außenbereich zu verbessern. Statt die Mehrweginformationen zu verwerfen, können wir sie konkret nutzen.

# Contents

# List of Figures

# List of Tables

# List of Algorithms

# 1. Introduction

The emitter localization problem has been a central research topic for many years [Cheng et al., 2012]; several current services rely on the possibility of extracting the position of a radio-frequency signal using a Wireless Sensor Network (WSN).

Contemporary outdoor communication systems are highly demanding for localization. A Non-Line Of Sight (NLOS) propagation takes place when there are obstructions (e.g., walls, vegetation, buildings, mountains) between the Transmitter (TX) and Receiver (RX). This causes multipath effects, an intrinsic urban channel impulse characteristic, which is challenging to eliminate.

A set of spatially separated sensors can measure the emitter radio-frequency signal and estimate its position. Among several parameters that can be measured, Time Of Arrival (ToA), Time-Difference Of Arrival (TDoA), and Received Signal Strength (RSS) provide information on the range between the source and sensors, while Direction Of Arrival (DoA) provides the source's bearings relative to the receivers [Wymeersch et al., 2009].

The NLOS condition causes significant positive biases in the estimated range. It is possible to use positioning fingerprints within a machine learning framework to enhance the precision of the estimation. However, this approach depends on large data sets extracted from the outdoor environment to build a feasible fingerprint model. Sometimes, this collection of real data is challenging or impractical to obtain.

There are several propagation tools for simulating the transmitter-receiver links (like Ray Tracing (RT)) that can generate a position fingerprint, defining a specific Channel Impulse Response (CIR) for each position in the environment. We propose a machine learning procedure that enhances the overall localization accuracy by combining the CIR from the RT simulation and real measurements from reference points to generate a mixed dataset.

It is possible to use the method inside the Cooperative Passive Coherent Location (CPCL) framework presented by [Thoma et al., 2019], where the localization problem is a Vehicle-to-Vehicle/Infrastructure (V2X) communication scenario, or even other applications in industry, transport, logistics, and security.

This initial chapter aims to describe the research motivations, delimiting the scope and defining the general terms to explain the problem addressed. We also present the state-of-the-art in the research of passive localization problems and list the main contributions to the related field.

## 1.1. Motivation

This thesis addresses the outdoor non-collaborative localization problem, where the multipath effect caused by non-line-of-sight propagation degrades the system performance. We assume a set of sensors deployed in a region with no synchronization between the emitter and the positioning system and no prior information on the signal.

The link between the target and the sensor in an outdoor scenario is usually affected by obstacles that reflect, diffract, or scatter the electromagnetic rays, causing the sensors to have imprecise information. In this case, a typical TDoA localization system presents a bias in the position processing.

There are several algorithms to mitigate this effect. Most of them try to identify if the signal received in each sensor is in a NLOS situation. In this case, the measurements affected by the multipath effect are assumed to be outliers and are removed from the positioning processing.

The initial motivation behind our proposal is to improve the precision of localization systems deployed in complex outdoor scenarios. The main idea is to provide additional information that could enhance the localization system performance by fusing ray-tracing simulation and signals from reference emitters.

Among the possible implementations, we have selected machine learning either in classification or a regression configuration to perform a positioning fingerprint approach. Although this idea could also be applied in indoor environments, we have limited the scope to non-collaborative outdoor scenarios.

This thesis addresses this issue by introducing a multipath effect fingerprint classification to enhance an existing localization system, giving an extra layer of information based on a hybrid approach that uses simulation data alongside real reference signals.

A typical machine learning setup relies on real measurements to build a model and tune the hyper-parameters, mapping the input-output dependence. We have introduced the idea to mix actual measurements and RT simulations to produce an initial data set, refining it with the real measurements as soon as they are available.

The localization problem can happen in a variety of areas, including car-to-car communication or autonomous driving, the Internet of the Things (IoT) and radio-frequency surveillance systems. There is always a trade-off among the possible sensor deployment positions, the desired system performance, and the area of interest. These constraints are even more relevant in an outdoor situation.

## 1.2. Scope and Definition

Several concepts and ideas presented in this thesis are also used in machine learning, localization, and signal processing. This section aims to clarify the project's scope and explain the terms used in the title. It is necessary to define the technical terminologies used through this text to describe the research scope appropriately. There are several taxonomies for localization problems; we adopted the one described in Figure 1.1.

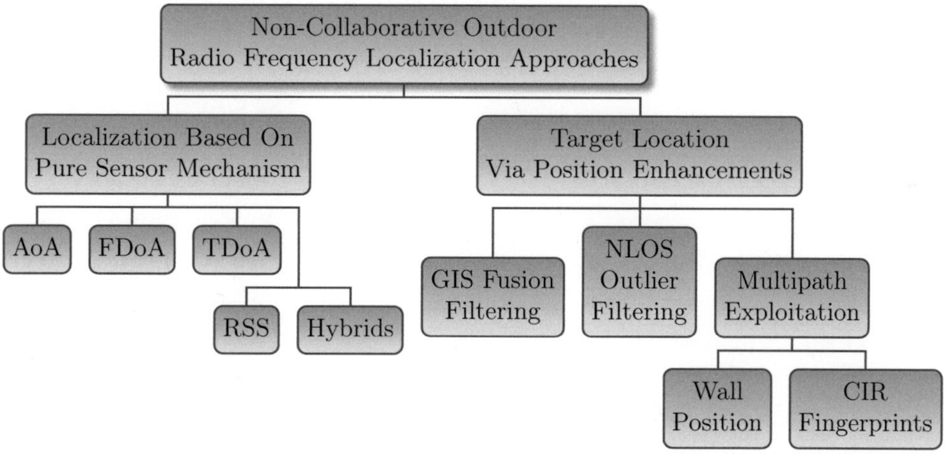


Figure 1.1.: Topologies of Localization [Zekavat and Buehrer, 2011].

Radio-frequency localization systems sometimes can count with the target collaboration, which means that either the signal structure or even some synchronization might be available in the received signal. In our approach, the sensor system does not have any information about the target signal, restricting our case to a "non-collaborative" localization problem.

In an outdoor scenario, this non-collaborative localization can use TDoA, Angle Of Arrival (AoA), Frequency-Difference Of Arrival (FDoA), RSS, a hybrid combination of these (Localization Based on Pure Sensor Mechanism), or different methods to enhance the position estimation (Localization via Position Enhancements) [Zekavat and Buehrer, 2011].

Regarding the enhancement strategies in [Vin et al., 2015], it is possible to use a Geographic Information System (GIS) to produce a kind of "context" filtering. Hence, if someone is locating a vehicle, the street information can be fused with the system position estimation to improve performance. It is possible to use statistical methods on the received signal to eliminate the corrupted measurements from a NLOS situation.

A multipath knowledge-aided method is a technique to explore some information hidden in the received signal related to the system environment. Some of these methods use the image theory to assign a virtual sensor and then use Simultaneous Localization and Mapping (SLAM) to enhance the position estimation. Systems can also exploit the multipath information if the locations of the reflecting walls are known and if the target specular components are resolvable and detectable.

We can use statistical methods and RT simulation to extract the specular components from the channel impulse response [Roivainen et al., 2016]. The system can assign each NLOS ray to the obstacles and reflection points in a given scenario.

The technique investigated herein exploits multipath of the environment and uses the knowledge of the actual direct path of the ray trajectory vector. The term "position fingerprints" refers to the situation where a specific combination of CIRs characterizes each place in the outdoor scenario.

There are two significant steps in using channel impulse information for the fingerprint approach. One is to perform the correct assignment between the features extracted from the receiver signal and the RT simulation. The other is determining how to use this mixture to build a data set that is representative enough to be used with the machine learning engine.

In short, the scope of this thesis is to use multipath exploitation based on channel impulse response fingerprints, enhancing a non-collaborative localization system in an outdoor scenario.

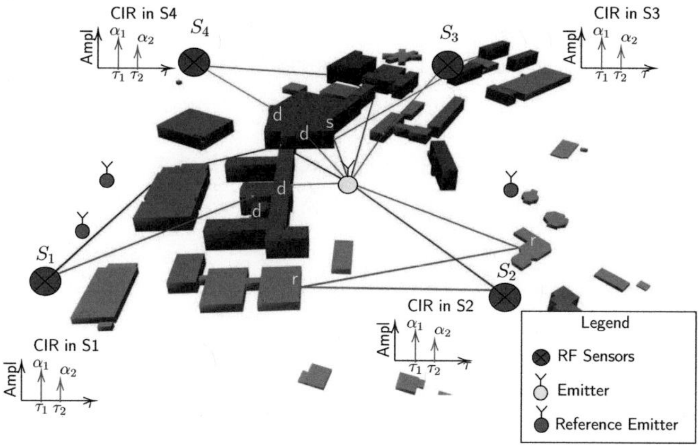


Figure 1.2.: Localization in Multipath Scenario.

# 1.3. Problems Addressed

This thesis discusses the general problem of radio-frequency localization in an outdoor scenario. The formulation of the problem starts with a typical localization setup by considering a radio-frequency emitter installed in an urban location where the signal received is blocked by the obstacles, as described in Figure 1.2. The figure shows that the incident rays can reflect, scatter, or diffract, depending on their position and angle of incidence. The channel impulse response has all this information, which is characteristic for each position.

Let us consider that the emitter position is a function of some specific parameter observed by the deployed sensors and defined by $\mathbf{f}(\mathbf{x})$. In a range-based system, the position estimation results from a nonlinear formulation process depending on the measurements observed by the localization system. The target location, defined as $\mathbf{f}(\tilde{\mathbf{x}})$, can use, among other methods, the Least Square (LS) or the Weighted Least Square (WLS) method.

Let $\mathbf{x}$ be the actual emitter position and $\tilde{\mathbf{x}}$ the estimated one, then the error function between them can be used as a parameter to evaluate the overall localization performance. The error function can be defined as:

$$\mathbf{e} = \mathbf{f}(\mathbf{x}) - \mathbf{f}(\tilde{\mathbf{x}}), \tag{1.1}$$

The total error in a localization system caused by multipath in a NLOS situation is model as follows:

$$\mathbf{e}_{TOTAL} = \mathbf{e}_{noise} + \mathbf{e}_{NLOS}. \tag{1.2}$$

There are statistical approaches to identify the measured signal by the sensors, whether received in Line Of Sight (LOS) or NLOS situation. However, the multipath effect is typical in an outdoor scenario, and the most realistic situation is unlikely to have a pure direct signal arriving at each sensor without reflections. These limitations show that developing methods to cope with systematic errors in outdoor localization applications is necessary.

Motivated by the possibility of using RT and machine learning, we build a position fingerprints structure that may help to increase the localization system performance.

We have proposed a technique to integrate measurements from reference emitters in the fingerprints structure, make a hybrid localization strategy where we theoretically use this information fusion to improve performance.

There are some isolated proposals [Alsheikh et al., 2014] to use machine learning and real measurements to build a position fingerprint that can enhance system performance. The RT simulation can give information, including the characteristics of obstacles and buildings, into channel impulse response characterization. Given the type of scenario, the scope delimitation, and the statement of the goals of this thesis, we can make the following problem formulation:

1. Is it possible to use multipath to improve position estimation systems?

2. Is it possible to use machine learning with datasets produced by a mixture of simulated signals and real calibration emitters?

3. Can the fusion of multipath fingerprints improve the position estimation of radio-frequency localization systems in an outdoor scenario?

Hence, we focus on the foundations to understand how the proposed method deals with multipath influences in the outdoor localization systems by considering this scope in the following chapters.

## 1.4. State of Art in Multipath Assisted Passive Localization

This section highlights the leading research in the field. It focuses on significant contributions to each research area described in detail in each chapter, discussing the literature foundations for each subtopic. The main idea here is to combine the fundamental investigations of radio-frequency localization systems, the multipath fingerprint, and machine learning in emitter localization.

It is necessary to conduct a literature survey to understand the approaches and proposed methods of the latest studies of multipath-assisted localization in outdoor scenarios. There is extensive literature on localizing 3D emitters based on measurements collected at sensors that can measure ToA, TDoA, FDoA, and DoA.

Although outdoor localization researches have several approaches, we choose the studies of [Li, 2006], [Liggins II et al., 2017] and [Reed et al., 1998] as our starting point in the literature review of NLOS effects on position processing.

[Reza, 2000] made an in-depth analysis of the localization in NLOS conditions and performed an investigation of a fusion approach for localization using ToA and TDoA. The study also includes a channel model formulated for evaluating position localization techniques within an NLOS cellular environment. The key point in this approach is modeling the effect of NLOS as a bias for TDoA measurements.

The work of [Porretta et al., 2004] presents a localization algorithm that makes use of a base station to locate cell phones in cellular networks. They explored the idea of using only one station as a sensor and multipath information to support position processing. They also showed a comparison of the proposed positioning technique compared to the well-known location method based on ToA measurements with more than three sensors.

[Algeier et al., 2010] has examined the idea of the multipath fingerprint, where he introduced a blind localization and tracking approach for mobile terminals in urban scenarios. The method used a track-before-detect algorithm that took advantage of multipath propagation information from primary scattering objects such as buildings. He also considered the idea of using a single observing station with the angle of the relative ToA of the multipath bound in a likelihood function that produces potential transmitter positions.

The researches conducted by [Landmann, 2008] addresses the problem of limits in experimental estimation and its performance in real-time propagation environments. He performed several measurement campaigns for channel modeling in an outdoor scenario using a channel sounder. The work showed a systematic way to define the limits in estimating specular components of outdoor situations.

The hybrid localization systems can use different types of measurements, as shown in [Steinberg and Bowman, 2017]. These systems can operate more effectively by fusing measurements and geographic information to refine the position estimation, expanding its geolocation capacity through the use of ToA and TDoA, as reported in [Hale, 2012].

[Mironov and Voroshilin, 2012] has investigated the dependency of the estimation accuracy on multipath components. The authors proposed to solve the problem by using signals reflected from the Earth's surface or reflecting objects, whose energies are higher than the direct signal energy, assuming that the reflected signals can be resolved at every receiving point using spatial processing.

The estimation framework obtained by [Taylor, 2013] has focused on RSS localization. It stated that the Geometric Dilution of Precision (GDOP) could provide a metric for evaluating a localization system's performance.

[Rehman et al., 2014] has examined the procedure of matching the radio frequency emitter's position with the signal feature that is characteristic from each point of the scenario. The study has provided an in-depth analysis of the research challenges of deploying a positioning system based on site-specific characteristics. According to the survey, It seems possible to assign a unique Radio Frequency (RF) fingerprint to a wireless device, essentially working as an identity in the same way as a biological fingerprint.

The fingerprint idea was further explored by [De Groot et al., 2014], who has provided an investigation of methods for tracking remote transmitters using a database generated by RT simulations. The positioning was accomplished by fusing the OpenStreetMap building data of the target area and a database including RSS, TDoA, and AoA fingerprints.

[Vin et al., 2014] has also proposed a radio frequency fingerprint approach for electronic surveillance in urban and suburban areas, where he suggested a framework to localize the emitter in complex outdoor NLOS conditions.

Further extensions on this topic were introduced by [Demissie, 2011], who has presented an algorithm for directly estimating the positions of multiple users of a mobile communication system in urban areas. This study aimed to deal with determining the delays and directions of arrivals of all multipath components.

The innovative idea of adding multipath information into localization processing has been proposed by [Schmitz et al., 2015], where an RT simulation has provided information about propagation paths. They have explained a multilateration method that was able to cope with multipath conditions in localization problems with no direct LOS to the target/sensor.

The investigation of [Gentner et al., 2016] has explained an algorithm that exploits multipath propagation for position estimation of mobile receivers. The authors have used recursive Bayesian filtering and the image theory to employ multipath components as signals emitted from virtual transmitters.

Localization systems can use statistical learning theory, as proved by [Brunato and Battiti, 2005]. The authors have presented an algorithm based on learning theory to solve the problem of determining the location fingerprinting of a wireless device. They have used a Support Vector Machine (SVM) to build a machine learning framework and compared the results with real measurements.

The work of [Ferguson et al., 2018] has proposed a convolutional neural network (CNNs) to improve the localization of sources of broadband acoustic radiated noise in a sound localization for the shallow water environment.

In the study of [Haidari et al., 2019], an RF localization scheme has been developed that can be applied in the absence of LOS signal reception based on the reflection phenomena. The proposed approach has used both RSS and DoA measurements with a Nearest Neighbor algorithm to estimate reflection parameters to obtain the location of the RF source.

A radio-frequency fingerprinting for 5G positioning was proposed by [Butt et al., 2020], where the authors suggested a machine learning assisted positioning method and evaluated their performance using simulations for an outdoor scenario in Lincoln Park, Chicago. The study also compares various Deep Learning models, showing that a positioning error in the range of 1-1.5m can be achieved.

The authors of [Al-Rashdan and Tahat, 2020] investigated localization based on the fingerprint approach for utilization in microcells urban environments. They tested several fingerprint machine learning for distributed massive multiple-input multiple-output (DM-MIMO) wireless systems configurations.

In short, the above literature review made clear to us that performance in outdoor RF localization schemes is severely deteriorated by NLOS situations. The problem has been addressed using various techniques, but multipath information in outdoor localization scenarios still has research gaps. Hence, it would be possible to use the channel information to enhance the performance of the system and the machine learning framework in multipath fingerprinting approaches.

## 1.5. Summary of Contribution

The aspects of this thesis are to attract researchers dealing with radio-frequency outdoor localization problems, not exclusively in mobile communication but, additionally, in passive localization, autonomous car guidance systems, IoT scenarios, and possibly future 5G deployments.

Throughout this research, we have developed and simulated several new concepts that contribute to state of the art in outdoor RF emitter tracking. To highlight some of them:

**TDoA Database Patterns:** We have built a technique for mapping position fingerprint patterns using ray-tracing and a TDoA database.

**RT and Actual Measurements Fusion:** We have developed an innovative method to aggregate RT simulations with real-time measurements to produce a data set for a machine learning fingerprints scheme.

**Position Estimation with CIR Fingerprints:** We have applied a Single Input Multiple Output (SIMO) Finite Impulse Response approach to combine the CIR fingerprints produced by RT and the Complex Baseband (CBB) signal arriving in each RF sensor.

**Neural Network using RT Multipath:** We have used a modified neural network localization, fusing localization fingerprints based in RT simulations with real measurements and applying the Inverse Covariance Intersection (ICI) scheme for enhancing outdoor localization systems.

**Fingerprints using Volume Cross-Correlation (VCC):** We also have included a synthetic position fingerprint method using VCC together with RT, fusing the estimations of a real localization system with the NLOS patterns and applying Random Forest Classification.

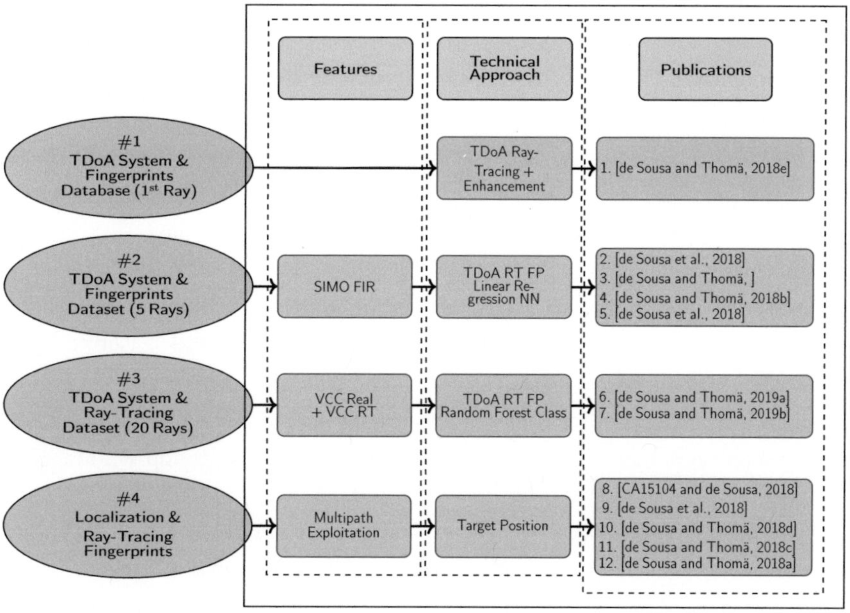


Figure 1.3.: Principal publications showing the research contributions.

The contribution of work covers different aspects of localization in outdoor scenarios and were presented in a number of publications, including two journal articles ( [de Sousa and Thomä, 2018b] and [de Sousa et al., 2021]), one special issue letter ( [de Sousa and Thomä, 2019a]), and several conference publications ( [CA15104 and de Sousa, 2018, de Sousa et al., 2018, de Sousa and Thomä, 2018e, de Sousa and Thomä, , de Sousa and Thomä, 2020, de Sousa and Thomä, 2018d, de Sousa and Thomä, 2018a, de Sousa and Thomä, 2018c, de Sousa and Thomä, 2019b]), as detailed in the following discussion.

Our principal contribution is the use of multipath information to enhance position estimation systems. Specifically, we considered the addition of scenario channel impulse information in the non-collaborative localization problem. We also proposed a RT simulation tool to form site-specific fingerprints to correlate NLOS rays to possible emitter points. Additionally, we used a classification machine learning engine to add real measurements into the simulated data set, enhancing the position prediction performance.

The relevance of this method is not only that it considers realistic outdoor scenarios but also that it combines RT fingerprints with real measurements. A small amount of previous work has tried to use machine learning in an indoor situation, and most of the localization strategies have considered a NLOS condition as an outlier in position processing.

This thesis summarizes several contributions to outdoor localization problems. In this work, the issue was addressed in progressive small steps, adding new findings and then taking the intermediate results as an input for the subsequent approach. Figure 1.3 shows not only the main proposed methods but also the link between the intermediate results and linking our specific research contributions with other published papers.

In the initial steps, we deployed a TDoA localization system in a campus-like scenario, which was harshly affected by a NLOS situation between the emitter and sensor nodes. A fingerprint database, produced with the "first ray" time differences, gave a set of possible TDoAs based on the sensor position and obstacle description. The final result has shown that the hybrid solution proposed gave an enhancement on TDoA system performance.

The work presented in [de Sousa and Thomä, 2018e] and [de Sousa et al., 2018] have shown the application of a RT propagation tool to build a geographic time difference fingerprint. The scheme has used mean and variance TDoA measurements taken from a localization system as a search parameter in the geo-database, showing likely positions of the emitter on the map. The idea was to improve the TDoA sensor performance to overcome multipath imprecision typical in the urban environment, where different rays are summed up at each sensor to correct the position estimation. This hybrid procedure depicted an actual improvement in a real scenario when almost all the sensors received only NLOS ray components.

As the next step, we produced results related to the studies in [de Sousa and Thomä, ], [de Sousa and Thomä, 2018a], and [de Sousa and Thomä, 2018b] and implemented the fingerprint idea by combining the five strongest rays in a regression neural network. This approach exhibited an enhancement in achieved results of the localization system, deployed in a suburban scenario with Dense Multi-path Component (DMC). We have also employed the concept of mixing the reference emitter signals to improve the accuracy of the machine learning engine.

The position fingerprint information produced by simulations together with calibration emitters provides a machine learning engine, which refines the target localization embedding all the reflection and diffraction in the propagation scenario.

RT fingerprints perform the target localization integrating multipath rays in the propagation scenario. Finally, we have matched the estimated NLOS components to a defined location. The proposed scheme is comparatively more accurate than other state-of-art techniques for position estimation.

In the third step, our investigations (published in [de Sousa and Thomä, 2019a], [de Sousa and Thomä, 2019b], and [de Sousa et al., 2021]) reported the fingerprinting strategy from the multipath data, produced by RT in the cross-correlation function in our machine learning framework. We have developed an artificial VCC strategy that matched the one provided by a TDoA system. The knowledge about the geometrical structure of the multipath propagation environment together with the classification random forest algorithm produced the data set of TDoA fingerprints for an outdoor scenario.

In the last stage of work, we have applied the fingerprinting idea and RT simulation to exploit the multipath information. We have proposed to locate an emitter by employing only one sensor and RT simulations to operate with the signal structure and the RT to improve the positioning processing. In [CA15104 and de Sousa, 2018], we have added some suggestions for the application of multipath exploitation in 5G wireless systems and IoT scenarios. [de Sousa and Thomä, 2018d] showed the combination of the RT simulation tool and the CIR estimation to perform the RF emitter location using a single sensor. In [de Sousa and Thomä, 2018c], we have exploited the possibility of matching the RT information localization procedure to locate Unmanned Aerial Vehicle (UAV) in an urban scenario and improved the TDoA performance in NLOS scenarios. Finally, in [de Sousa and Thomä, 2020] we have suggested applying the multipath fingerprints in a tactical scenario, where a navy frigate is in Ipanema, Rio de Janeiro, Brazil, and tries to track an RF emitter target in a hypothetical counterinsurgency situation.

Figure 1.4 shows the complete framework achieved after each step, presented in papers and conferences. The idea is applied in a typical TDoA system deployed in a NLOS scenario, where the multipath propagation influences the position estimation. At the same time, a data set of the CIR is developed using the RT simulation and measurements of the calibration points. The machine learning framework is used to produce the position fingerprints model that assigns the CIR features to each point in the scenario, which is then fused with the real-time estimation, producing an enhancement in position estimation.

In summary, our contribution to the research field is a hybrid procedure combining real-time measurements and simulation. For this purpose, the RT and machine learning improve the radio-frequency localization performance in an outdoor scenario, where it is difficult or even not feasible to build a data set solely based on real-time measurements.

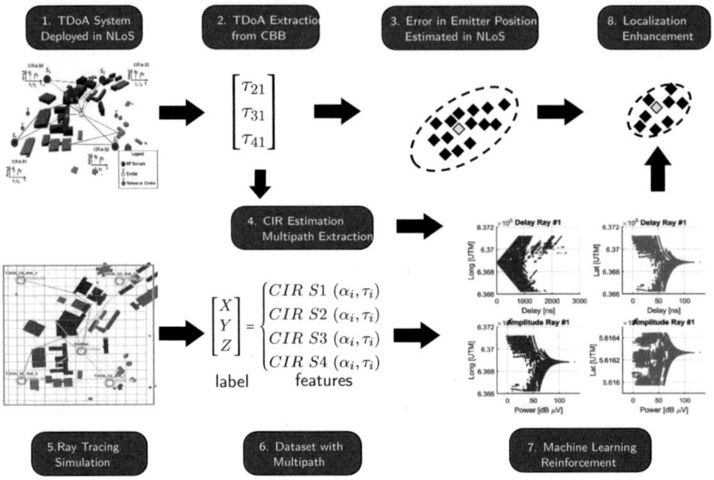


Figure 1.4.: Multipath Exploitation using Ray-Tracing Fingerprints and Machine Learning.

## 1.6. Thesis Outline

In this chapter, we have discussed the motivation for this work, the scope and definitions used, the problem addressed, the current status of research, and a summary of the main results. We have divided the rest of the thesis into two parts; Part I (Chapters 2, 3, 4, and 5) presents the fundamentals of RF emitter position and multipath exploitation with RT and Part II (Chapters 6 and 7) describes the applications and the validation scenario of this work.

Chapter 2 describes the fundamentals of RF emitter localization systems, shows processing and formulation aspects of position estimation. There is a mathematical foundation of the non-collaborative passive localization, where we present the relationship between the multipath and the bias in the target position estimation.

Chapter 3 illustrates the main aspects of channel impulse estimation on the proposed fingerprint approach, comprised of path loss, multipath fading, and shadowing effects. These parameters in the link between the target and receiver are the foundation of our multipath exploitation localization technique.

In Chapter 4, we present the fundamentals of the multipath characterization using RT, which is a propagation tool employed for communication system deployment. We illustrate the key aspects related to the NLOS extraction to build a data set that incorporates the most significant effects on the channel to define each possible position.

The machine learning system is described in Chapter 5, where we explain the fundamental differences between regression and the classification setup incorporated in the estimation models. It also gives a summary of the machine learning tools and related terminologies used in our model.

In Chapter 6, the complete framework of the multipath based on RT is presented, where we correlate the essential foundations explored in the previous sections and give a detailed description of the fingerprints applied either in the CIR or in the cross-correlation function. This chapter also shows the well-known family of *random forests* methods employed to implement our machine learning system.

Chapter 7 describes various field experiments where the proposed technique is applied to validate our approach. We discuss the measurement campaigns conducted on the campus of the Technische Universität Ilmenau (TUI) and in a suburban environment in the city.

# Part I

# Fundamentals of RF Emitter Position with Ray-tracing Multipath Exploitation

# 2. RF Emitter Position Framework

This chapter aims to explain the fundamentals of emitter localization systems and describe important positioning estimation aspects. It gives an overview of the fundamentals to determine the position of a radio-frequency target using a set of distributed sensors. It is structured into three main sections, starting with a general foundation of the localization problem, possible implementations, including algorithms, and constraints of practical systems.

A set of distributed sensors with known location can determine the position of a radio frequency emitter using the signal received in a collaborative or in a non-collaborative way. In a non-collaborative way, the localization system uses only the signal received without synchronization with the transmitter. The source localization can use the receiver signal that arrives in each sensor or the knowledge from the environment.

A localization system can use distance or angle information to estimate the emitter position. The localization can extract Time Of Arrival (ToA), Time-Difference Of Arrival (TDoA), Received Signal Strength (RSS) to estimate the emitter position using relative distances; or it can use the Angle Of Arrival (AoA) in each sensor to extract the relative bearings to the receivers.

TDoA is a high-precision emitter location technique that is frequently used to track noncooperative emitting targets over time. Its basic premise is to measure the time difference of arrival at several sensor locations and use those measurements to estimate the emitter position.

The relation between the emitter position and the TDoA is nonlinear, demanding more elaborate algorithms for performing the estimation. The first class of solution can operate directly using the relationship between the source and the measurements, namely, Nonlinear Least Squares (NLS) and Maximum Likehood (ML) estimators. Another class uses linearization approaches on the equations, namely, Linear Least Squares (LLS), Weighted Linear Least Squares (WLLS), and subspace approach.

The analysis starts with the assumption that there are no outliers in the measurement with a direct path link between transmitter and receiver. Later we show that if Non-Line Of Sight (NLOS) occurs, they produce a positive bias in the corresponding distance information, which is the primary source of errors is multipath propagation.

It might be possible to extend the application of the proposed research on different localization methods, such as AoA, Frequency-Difference Of Arrival (FDoA) and RSS. However, range systems usually are based on single-antenna devices, with small size, weight, and power (SWAP) [Pourhomayoun and Fowler, 2014], which turns to be smaller and more straightforward than bulky direction-finders used in AoA systems. This work focuses on TDoA systems and their formulation.

To fully understand position estimation, we can split the processing into three main steps: the mathematical formulation of the data model, the parameter extraction, and the position estimation algorithms. In Section 2, there is a geometrical analysis of the non-collaborative passive localization, with the derivation of the fundamental equations, using a generic matrix format.

Throughout the chapter, we perform several simulations to support the explanations of the concepts. We point out the key aspects essential to understand our contribution and connect with the other parts of this thesis. The simulations use a typical localization algorithm and include noise and multipath effects on the localization system.

Finally, we present a discussion about the constraint and design trade-off, analyzing the multipath effect on the position estimation.

## 2.1. Motivation and Related Work

This section discusses the investigations related to the outdoor localization problem, where we have selected those necessary to understand the contributions produced. Localization is a classical estimation problem in the signal processing community, and the position estimation has been a central issue in several fields of applications such as radar, sonar, telecommunication, and [Carter, 1993], telecommunications [Huang and Benesty, 2007, Jagoe, 2003] and wireless sensor networks [Ilyas and Mahgoub, 2004, Stojmenovic, 2005].

[CA15104 and de Sousa, 2018] shows that, in the context of the Internet of the Things (IoT), it is possible to incorporate low power sensor localization nodes in small, inexpensive devices that can collect information from the surroundings and collaboratively communicate among them. It is an enabling approach for remote monitoring and control applications, where node positioning is a fundamental issue for network management.

In [Torrieri, 1984] and [Friedlander, 1987], we can find a precise derivation of the position algorithms and the time difference processing using the cross-correlation of the sensor signals. There is also a description of the position performance based on concentration ellipse, the Circular Error Probability (CEP), and the Geometric Dilution of Precision (GDOP) and a description of Doppler effects. [Mironovova and Havlis, 2011] derives a GDOP coefficient, useful for using as a localization performance tool.

[Stein, 1993] performed an analysis of the application of maximum likelihood ML estimation on noise corrupted Gaussian mixtures, where the author derived the Complex Ambiguity Function (CAF) of narrow-band signals received by different sensors. [Ho and Chan, 1997] offered an extension of the position estimation, also including the relative motions between an emitter and receivers, proposing an FDoA combined with TDoA measurements.

In [Weiss and Stein, 1987], the authors proposed a classical approach to estimate the time delay of a random signal observed at two spatially separated sensors. The method operates on complex envelopes, including real measurements with a real-time hardware implementation of the algorithm; the results showed a close agreement with the theoretical prediction.

[Reza, 2000] exposed a TDoA localization scheme using the maximum likelihood ML estimation, including the noise into the formulation. He analyzed the nonconvex nature of ML function and the difficulty to find the solution when including the sensor location errors and NLOS effects.

In [Ho et al., 2007], there is a formulation of the impact in the inaccurate knowledge of sensor position for locating a moving source using time-difference-of-arrival TDoA and frequency-difference-of-arrival FDoA measurements in the presence of random noise.

While [Fowler and Hu, 2008] explored the hybrid TDoA / FDoA approach, explaining the significant differences between the formulation for the acoustic and the radio-frequency signals, [Musicki et al., 2009] studied the recursive tracking of one mobile emitter using a sequence of time difference of arrival TDoA and frequency difference of arrival FDoA. [Cho et al., 2012a] also explored the hybrid TDoA / FDoA approach, including a comparison of the particle filter and the extended Kalman filter in the position estimation. Another relevant approach is the findings of [Ho, 2012] to reduce the bias of the source location estimation, corrupted by significant noise.

[Tas, 2012] presents relevant work related to TDoA geolocalization, sharing the GDOP formulation that was enhanced and used in our studies as a figure of merit to evaluate the enhancement in the positioning systems.

Some non-linear estimation methods are using a Weighted Least Square (WLS) for source localization problem, either using TDoA / FDoA , as described by [Qu et al., 2017], [Noroozi et al., 2017] and [Cheney and Given, 2017]. [Zou et al., 2017] showed a maximum likelihood ML estimator for non-linear cases as well, which was further improved by [Zhu et al., 2018] and [Dexiu et al., 2018].

There are several possible applications for localization services in the cellphone standards. The work of [Zekavat and Buehrer, 2011] provides an overview of the evolution of the various localization methods. [El Gemayel et al., 2013] showed a low-cost TDoA systems based on software-defined radios; and in [El Gemayel et al., 2013]. Furthermore, [Degen et al., 2014] explained a passive non-cooperative localization and tracking of mobile terminals in urban scenarios called blind mobile localization (BML), [Degen et al., 2014].

Improvement in the TDoA processing was not the focus of our work, but it is necessary to understand how an entire system operates and how it deals with the multipath problem. The formulation of [Fokin et al., 2018] delivered valuable insights because they proposed a three-stage time difference of arrival in non-line-of-sight (NLOS) conditions.

A classical way to deal with NLOS is to consider these measurements as outliers, [Compagnoni et al., 2017a], where the author proposes outlier removal procedures, enabling the identification of spurious measurements within a set. He performed further analyses, presented in [Compagnoni et al., 2017b], showing the source localization as a denoising problem of TDoA.

The studies of [Wang et al., 2016] addressed the TDoA problem under mixed line-of-sight (LOS)/non-LOS (NLOS) conditions. [Yin, 2014] provided a more realistic model for a multipath environment, including two new classes of localization algorithms to estimate the positions and measurement error statistics jointly. [Lee et al., 2017] applied a Kalman filter-based NLOS section identification method for iteratively estimate emitter location using the recursive weighted least square algorithm. [Velasco et al., 2016] suggested admitting a closed-form solution for TDoA, assuming that the Gaussian noise contaminates the measurements, modeling the case of having missing data.

The studies listed in this chapter aim to provide state-of-the-art radio-frequency outdoor localization problems. Typically, publications have focused more on signal processing and optimizing the positioning of sensors, evaluating and comparing different implementations. In this sense, the sound disturbances considered are, as a rule, modeled as Gaussian. Few studies make a comparative study with real measures.

Our contribution to the field of research lies in the fact that the analysis made from theoretical formulations reaches the practical effects of an existing location system. Thus, we investigate the multipath effects, the bandwidth restrictions, and the lack of perfect synchronism between stations, presenting a realistic idea of existing limitations and their impact on outdoor location.

In summary, we have pointed out the leading related research for localization including, data model, algorithms, and signal processing focusing on TDoA, adding the drawbacks of the multipath effects from the NLOS on the position estimation. The ideas and results exposed herein are necessary to understand the concepts detailed in the following sections. Finally, we present a discussion about the constraint and design trade-off, analyzing the multipath effect on the position estimation.

## 2.2. Data-Model for Localization using TDOA

We start the formulation using the derivation presented in [Carter, 1987] and [Weiss and Stein, 1987], where we assume that there is a clock synchronization across all receivers. Mathematically, the TDoA scheme is formulated as illustrated in Figure 2.1, where a target emits a radio-frequency signal at the unknown time $t_0$ and the $l^{th}$ sensor receives it at time $t_l, l = 1, 2, \ldots, L$ with $L \geq 3$ [Stein, 1981].

Assuming that the general sensor-pair time differences or delays are defined as:

$$\begin{aligned} t_{k,l} = \quad & (t_k - t_0) - (t_l - t_0) \\ & t_k - t_l, k, l = 1, 2 \ldots, L \quad (k > l) \end{aligned} \tag{2.1}$$

If we use $L$ sensors, there are $L(L-1)/2$ distinct delays, some works [Khalaf-Allah, 2014] and [Apolinário et al., 2019] use the combinations of possible TDoAs, in a so-called extended localization scheme to achieve superior localization performance.

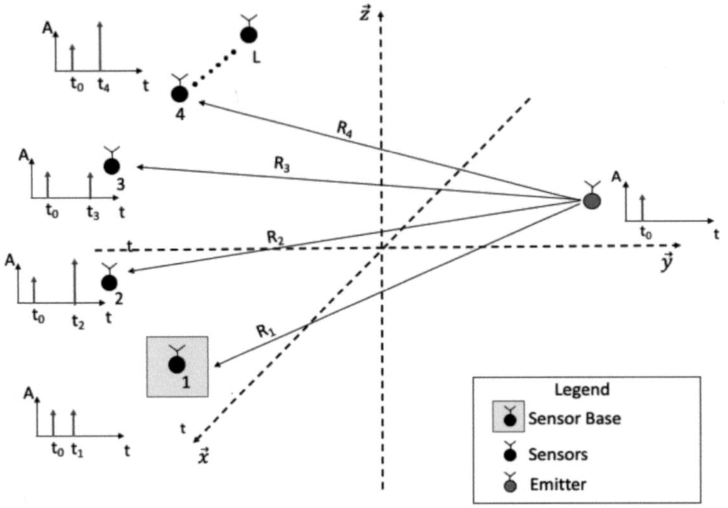


Figure 2.1.: Framework of Machine Learning Implementation.

After that, we will restrict our analysis to a traditional approach where one sensor acts as a *Fusion Node*, where all the measurements are refereed. Assuming Sensor 1 as a reference, the distinct time-differences are $t_{i,1}\, l = 2, 3, \ldots, L$, and there are $(L-1)$ distinct measurements, with one central fusion node.

Geometrically, each noise-free time-difference defines a branch of a hyperbola in 2D or a hyperbolic surface in 3D. A hyperbola is geometrically a set of points, where the absolute difference of the distances to two fixed points (the foci) is constant. Then, the target location can be obtained by the intersection of at least two hyperbolas [Zekavat and Buehrer, 2011].

In the minimum case with three sensors, we have three distinct time differences in 2D space. The emitter location could be where the two TDOA hyperbolas intersect, corresponding to two possible source locations. Depending on the geometry (sensor locations) and possible noisy measurements, they may not cross. They could even have multiple intersections, so the intersection search is not an effective way to solve the problem.

With more than three sensors, it is possible to convert the noisy TDoA into a set of equations, where the source position can be determined numerically using minimum distance search optimization criteria.

Let $x = [x, y, z]^T$ be the unknown source position and $x_l = [x_l, y_l, z_l]^T$ be the known coordinates of the $l^{th}$ sensor, $l = 1, 2, \ldots, L$, where $L \geq 3$ is the number of receivers. The distance between the emitter and the lth sensor is denoted by $d_l$:

$$d_l = \|x - x\|_2 = \sqrt{(x - x_l)^2 + (y - y_l)^2 + (z - z_l)^2}, \qquad l = 1, 2, \ldots\ldots., L. \tag{2.2}$$

The source emits a signal at time $t_0$ and the $l^{th}$ sensor receives it at time $t_l$; that is, $t_l$ is the time of arrival in each sensor, that can be converted into distance. We have a simple relationship between $t_l$ and $d_l$ :

$$t_l = \frac{d_l}{c}, \qquad l = 1, 2...., L. \tag{2.3}$$

The difference of time arriving at each sensor $TDoA = t_{L,l}$ is related with the range difference of arrival $r_{TDoA,l}$ by:

$$t_{L,l} = \frac{r_{TDoA,l}}{c}, \qquad l = 1, 2...., L. \tag{2.4}$$

The terms time-difference and range differences are interchangeable, their relation being the speed-of-light. If the actual range differences is corrupted by the noise $(n)$, they are modeled:

$$r_{TDoA,l} = d_{l,1} + n_{TDoA,l}, \qquad l = 2, 3, ....., L; \tag{2.5}$$

where

$$d_{l,1} = d_l - d_1, \tag{2.6}$$

and $n_{TDoA,l}$ is the range difference error in $r_{TDoA,l}$, which is proportional to the time delays $t_{l,1}$.

Repeating the same procedure of Equations (2.2) - (2.6), including all distinct time-differences, with Sensor 1 as the base, we build a TDoA signal model in matrix form as:

$$\boldsymbol{r}_{TDoA} = \mathbf{f}_{TDoA}(\boldsymbol{x}) + \mathbf{n}_{TDoA}, \tag{2.7}$$

where,

$$\boldsymbol{r}_{TDoA} = \left[r_{TD0A,2} r_{TDoA,3}, \ldots, r_{TDoA,L}\right]^T, \tag{2.8}$$

$$\mathbf{n}_{TDoA} = \left[n_{TDoA,2} n_{TDoA,3}, \ldots, n_{TDoA,L}\right]^T, \tag{2.9}$$

and

$$\mathbf{f}_{TDoA}(\boldsymbol{x}) = \boldsymbol{d}_l$$
$$= \begin{bmatrix} \sqrt{(x-x_2)^2+(y-y_2)^2+(z-z_2)^2} - \sqrt{(x-x_1)^2+(y-y_1)^2+(z-z_1)^2} \\ \sqrt{(x-x_3)^2+(y-y_3)^2+(z-z_3)^2} - \sqrt{(x-x_1)^2+(y-y_1)^2+(z-z_1)^2} \\ \vdots \\ \vdots \\ \sqrt{(x-x_L)^2+(y-y_L)^2+(z-z_L)^2} - \sqrt{(x-x_1)^2+(y-y_1)^2+(z-z_1)^2} \end{bmatrix}. \tag{2.10}$$

The function $\mathbf{f}_{TDoA}(\boldsymbol{x})$ operates the time-difference matrix, where the localization problem using TDoA measurements is defined as an estimation of $\boldsymbol{x}$ given the set of range-differences in matrix form $\boldsymbol{r}_{TDoA}$.

Using a Gaussian with zero-mean to model the noise ($\mathbf{n}_{TDoA}$), the corresponding Probability Density Function (PDF) of $\boldsymbol{r}_{TDoA}$ will be defined as $p(\boldsymbol{r}_{TDoA})$, with the structure given by Equation (2.11), where $\mathbf{C}_{TDoA}$ is the covariance matrix. The noise can be also represented by the notation $\boldsymbol{r}_{TDoA} \sim N(\boldsymbol{r}_1, C_{TDoA})$.

$$p(\boldsymbol{r}_{TDoA}) = \frac{1}{(2\pi)^{(L-1)/2}\left|\mathbf{C}_{TDoA}\right|^{1/2}} exp(-\frac{1}{2}(\boldsymbol{r}_{TDoA} - \boldsymbol{r})^T \times \mathbf{C}_{TDoA}^{-1}(\boldsymbol{r}_{TDoA} - \boldsymbol{r}); \tag{2.11}$$

where $\mathbf{C}_{TDoA}$ is the covariance matrix for $\boldsymbol{r}_{TDoA}$, corresponding to

$$\begin{aligned} \mathbf{C}_{TDoA} &= E\{(\boldsymbol{r}_{TDoA} - \boldsymbol{r})(\boldsymbol{r}_{TDoA} - \boldsymbol{r})^T\} \\ &= E\left\{\mathbf{n}_{TDoA}\mathbf{n}_{TDoA}^T\right\}. \end{aligned} \tag{2.12}$$

The correlation matrix of the signals measured by $L$ TDoA sensors is $\mathbf{C}_{TDoA}$; then the time-differences are taken from the $N = 2$ distinct pairs. All TDoA measurements are extracted via a correlation between the signals received in the different sensors and the base-sensor.

The covariance matrix assumes the following structure:

$$\mathbf{C}_{TDoA} = c^2 \begin{bmatrix} \sigma_1^2 + \sigma_N^2 & \sigma_N^2 & \cdots & \sigma_N^2 \\ \sigma_N^2 & \sigma_2^2 + \sigma_N^2 & \cdots & \\ \vdots & & \ddots & \vdots \\ \sigma_1^2 & \cdots & & \sigma_{N-1}^2 + \sigma_N^2 \end{bmatrix}. \tag{2.13}$$

The term $\sigma_i^2$ is the variance in TDoA measurements, and term $c$ is the speed of light. The covariance matrix $\mathbf{C}_{TDoA}$ is not diagonal, because we have taken the first sensor as the base for all TDoAs, then $\mathbf{n}_{TDoA,l}$, $l = 2, 3, \ldots, L$ are correlated. We can use the approximation given by [Zekavat and Buehrer, 2011], assuming $\boldsymbol{r}_{TDoA} \sim N(\boldsymbol{d}_l, \mathbf{C}_{TDoA})$.

## 2.3. Algorithms for localization using TDoA

The basic idea in this subsection is to give an overview of the existing classical methods for performing localization using a time-difference measurement. Consequently, it is appropriate to show the formulation of these methods before examining more practical aspects of a real localization system.

We have focused on linear and nonlinear procedures, but other approaches, like Particle Filters (PF) or Extended Kalman Filter (EKF). There is an in-depth comprehensive derivation, showing other implementation in [Torrieri, 1984] and [Zekavat and Buehrer, 2011].

For our investigation, it is pertinent to understand how the multipath and the error could affect the position estimation based on TDoA. The nonlinear methods try to use a direct estimate, solving Equation (2.4) directly. They use the value of $\boldsymbol{x}$ that minimizes the Least Square (LS) or WLS error, used as a cost function. It is possible to make non-linear algorithms using the Gauss-Newton, Newton-Raphson or Steepest-descent search method. The estimation $\tilde{\boldsymbol{x}} = [\tilde{x}\ \tilde{y}\ \tilde{z}]^T$ is the optimization variable that minimizes the error, to be defined here as a non-linear least squared estimator, as follows:

$$\mathbf{e}_{nonlinear} = \boldsymbol{r}_{TDoA} - \mathbf{f}_{TDoA}(\tilde{\boldsymbol{x}}). \tag{2.14}$$

The linear approach in Equation (2.14) relates to the position and distances differences and can be linearized, forming a system with a set of linear equation in $\boldsymbol{x}$

$$\mathbf{b} = \mathbf{A}\boldsymbol{x} + \mathbf{q}, \tag{2.15}$$

where $\mathbf{b}$ and $\mathbf{A}$ are known, and $\mathbf{q}$ is the noise vector. Based on Equation (2.15), we construct

$$\mathbf{e}_{linear} = \mathbf{b} - \mathbf{A}\tilde{\boldsymbol{x}}. \tag{2.16}$$

The linear algorithms use Equation (2.16) to perform linear LS, weighted linear LS or subspace estimation, as presented by [Chan et al., 2008] and [Zekavat and Buehrer, 2011].

### 2.3.1. Nonlinear Algorithms

The nonlinear methods estimate a target location directly from Equations (2.2), using either the Non-Linear Leas Squared (NLS) or ML.These schemes may not achieve a global convergence because their optimization cost functions can reach a local maximum or minimum.

Usually, if the noise characteristics are unknown, the nonlinear LS is used because they are simple to implement. It is also possible the ML estimator, which uses noise covariance information as a weighted LS, as explained in [Rao et al., 1973], achieving the Cramér-Rao Lower Bound (CRLB) limit asymptotically.

The estimation using the NLS is equal to $\tilde{x}$, meaning that it is the smallest value of $J_{NLS,TDoA}(\tilde{x})$; that is,

$$\hat{x} = \arg\min_{\hat{x}} J_{NLS,TDoA}(\tilde{x}). \tag{2.17}$$

The non-linear cost function for the TDoA positioning, is defined as $J_{NLS,TDoA}(\tilde{\boldsymbol{x}})$, and has the follwoing structure:

$$
\begin{aligned}
\mathbf{J}_{NLS,TDoA}(\tilde{\boldsymbol{x}}) &= \sum_{l=2}^{L}\Big(r_{TDoA,l} - \sqrt{(\tilde{x}-x_l)^2 + (\tilde{y}-y_l)^2 + (\tilde{z}-z_l)^2} \\
&\qquad + \sqrt{(\tilde{x}-x_1)^2 + (\tilde{y}-y_1)^2 + (\tilde{z}-z_1)^2}\Big), \\
&= \left(\boldsymbol{r}_{TDoA} - \mathbf{f}_{TDoA}(\tilde{\boldsymbol{x}})\right)^T \left(\boldsymbol{r}_{TDoA} - \mathbf{f}_{TDoA}(\tilde{\boldsymbol{x}})\right).
\end{aligned} \tag{2.18}
$$

The nonlinear relation between distance and position $\hat{x}$ is challenging to obtain. The recursive optimization search can end up in a local minimum instead of the global solution in the 2-D surface of $J_{NLS,TOA}(\tilde{x})$.

When the noise distribution is known, the ML algorithm maximizes the PDF of the TDoA measurements to obtain a better emitter position estimation. [Stein, 1993] explained that the maximization of Equation (2.8) corresponds to a weighted version of the NLS scheme if the noise is zero-mean Gaussian distributed,

Using the Equation (2.8), the ML cost function for TDoA positioning, denoted by:

$$
\mathbf{J}_{ML,TDoA}(\tilde{\boldsymbol{x}}) = \left(\boldsymbol{r}_{TDoA} - \mathbf{f}_{TDoA}(\tilde{\boldsymbol{x}})\right)^T \mathbf{C}^{-1} \left(\boldsymbol{r}_{TDoA} - \mathbf{f}_{TDoA}(\tilde{\boldsymbol{x}})\right) \tag{2.19}
$$

The position estimation with the ML is equal to $\tilde{\boldsymbol{x}}$, meaning also the smallest value of $J_{ML,TDoA}(\tilde{\boldsymbol{x}})$.

$$
\hat{\boldsymbol{x}} = \arg\min_{\tilde{\boldsymbol{x}}} J_{ML,TDoA}(\tilde{\boldsymbol{x}}) \tag{2.20}
$$

It is possible to find $\hat{\boldsymbol{x}}$ by finding the global minimum in the 2-D surface of $J_{NLS,TOA}(\tilde{\boldsymbol{x}})$. As described in [Zekavat and Buehrer, 2011], there are two schemes for solving Equation (2.20), one of them is based on global exploration using grid search with a genetic algorithm or particle swarm optimization.

Another one is based on a local search, with iterative algorithms using Newton-Raphson, Gauss-Newton, or steepest descent methods. Both Newton-Raphson and Gauss-Newton methods provide fast convergence; it is necessary to invert a matrix in the former. The latter is simpler because is based on the second-order differentiation of the $J_{NLS,TOA}(\tilde{\boldsymbol{x}})$ parameter.

### 2.3.2. Linear Algorithms

The main concept of this type of localization algorithm is to convert the nonlinear distance-position Equations (2.2), (2.4) and (2.7) in a set of linear equations, solving the related system.

The Global solution is obtained if we assume that the errors in the range-differences are small and with the zero-mean distribution. There are several linear methods, like Linear Least Square (LLS), Weighted Linear Least Square (WLLS), and subspace decomposition, where the WLLS provides higher localization accuracy.

However, to apply these methods, it is necessary to know the mean and covariance of the errors in the linear equations. We can extract the values of the weight terms if the noise structure is either provided or estimated. An eigenvalue decomposition (EVD) process generates the source estimation, relating the $\boldsymbol{x}$ with the range differences.

Using an intermediate variable ($m$) in the linearization process of the derivation, we can rewrite Equation (2.2) and Equation (2.3) as:

$$\begin{aligned} r_{TDOA,l}+&\sqrt{(x-x_1)^2+(y-y_1)^2(z-z_1)^2} = \\ &\sqrt{(x-x_l)^2+(y-y_l)^2+(z-z_l)^2}+n_{TDOA,l}^2, \quad l=2,3,\ldots,L. \end{aligned} \tag{2.21}$$

$m_{TDOA,l}$ is a variable with the modified noise component and a dummy variable $R_1$:

$$m_{TDOA,l} = n_{TDOA,l}^2 + 2n_{TDOA,l}\sqrt{(x-x_1)^2+(y-y_1)^2+(z-z_1)^2}. \tag{2.22}$$

$$R_1 = \sqrt{(x-x_1)^2+(y-y_1)^2+(z-z_1)^2}. \tag{2.23}$$

Squaring both sides of Equation (2.21), and using Equation (2.22) and (2.23), we obtain:

$$\begin{aligned}2(x_1-x_l)(x-x_1)+2(y_1-y_l)(y-y_1)2(z_1-z_l)(z-z_1)\\ -2r_{TDoA,l}R_1+m_{TDoA,l}\\ =r^2_{TDoA,l}-(x_1-x_1)^2-(y_1-y_1)^2-(z_1-z_1)^2, \qquad l=2,3,..,L.\end{aligned} \tag{2.24}$$

If we define:

$$\mathbf{A}=2\begin{bmatrix} x_1-x_2 & y_1-y_2 & z_1-z_2 & -r_{TDOA,2}\\ x_1-x_3 & y_1-y_3 & z_1-z_3 & -r_{TDOA,3}\\ \vdots & \vdots & \vdots & \vdots\\ x_1-x_L & y_1-y_L & z_1-z_L & -r_{TDOA,L}\end{bmatrix}, \tag{2.25}$$

$$\boldsymbol{\theta}=\begin{bmatrix} x-x_1 & y-y_1 & z-z_1 & R_1\end{bmatrix}^T. \tag{2.26}$$

$$\mathbf{q}=[m_{TDOA,2},\ m_{TDOA,3},\ \ldots, m_{TDOA,L}]^T, \tag{2.27}$$

and,

$$\mathbf{b}=\begin{bmatrix} r_{TDOA,2} & -(x_1-x_2)^2 & -(y_1-y_2)^2 & -(z_1-z_2)^2\\ r_{TDOA,3} & -(x_1-x_3)^2 & -(y_1-y_3)^2 & -(y_3-y_3)^2\\ \vdots & \vdots & \vdots & \vdots\\ \vdots & \vdots & \vdots & \vdots\\ r_{TDOA,L} & -(x_1-x_L)^2 & -(y_1-y_L)^2 & -(z_1-z_L)^2\end{bmatrix}. \tag{2.28}$$

The Equation (2.24) in matrix becomes:

$$\mathbf{A}\boldsymbol{\theta}+\mathbf{q}=\mathbf{b}. \tag{2.29}$$

The sensor positions are used to build the $(L-1)\times 4$ matrix $\mathbf{A}$ and the TDoA observations in vector $\boldsymbol{r}_{TDoA}$ is transformed in $(L-1)\times 4$ vector $\mathbf{b}$.

The emitter location is given by the estimated matrix $\boldsymbol{\theta}$. If the noise in each sensor is sufficiently small ($m_{TDoA,l}$), the terms $\{n^2_{TDOA,l}\}$ can be neglected, so $E[\mathbf{q}]\approx 0$.

$$\mathbf{A}\boldsymbol{\theta} \approx \mathbf{b} \tag{2.30}$$

$$\begin{aligned} J_{LLS,TDoA}(\tilde{\boldsymbol{\theta}}) &= (\mathbf{A}\tilde{\boldsymbol{\theta}} - \mathbf{b})^T(\mathbf{A}\tilde{\boldsymbol{\theta}} - \mathbf{b}) \\ &= \tilde{\boldsymbol{\theta}}^T\mathbf{A}^T\mathbf{A}\tilde{\boldsymbol{\theta}} - 2\tilde{\boldsymbol{\theta}}^T\mathbf{A}^T\mathbf{b} + \mathbf{b}^T\mathbf{b}, \end{aligned} \tag{2.31}$$

The term $J_{LLS,TDoA}(\tilde{\boldsymbol{\theta}})$ is a quadratic function in $\tilde{\boldsymbol{\theta}}$, where the single minimum is:

$$\hat{\boldsymbol{\theta}} = \arg\min_{\tilde{\theta}} J_{LLS,TDoA}(\tilde{\boldsymbol{\theta}}) \tag{2.32}$$

To find the minimum, we differentiate Equation (2.32) with respect to $\tilde{\boldsymbol{\theta}}$, setting the result to zero:

$$\begin{aligned} \frac{\partial J_{LLS,TDoA}(\tilde{\boldsymbol{\theta}})}{\partial \tilde{\boldsymbol{\theta}}}|_{\tilde{\boldsymbol{\theta}}=\boldsymbol{\theta}} &= \mathbf{0} \\ 2\mathbf{A}^T\mathbf{A}\boldsymbol{\theta} - 2\mathbf{A}^T\mathbf{b} &= \mathbf{0} \\ \mathbf{A}^T\mathbf{A}\boldsymbol{\theta} &= \mathbf{A}^T\mathbf{b} \\ \boldsymbol{\theta} &= (\mathbf{A}^T\mathbf{A})^{-1}\mathbf{A}^T\mathbf{b} \end{aligned} \tag{2.33}$$

The emitter position estimation $[x, y, z]$ using LLS is simply extracted from the first, the second and third entries of $\tilde{\theta}$ that is

$$\hat{\boldsymbol{x}} = [[\hat{\boldsymbol{\theta}}]_1 + x_1 \quad [\hat{\boldsymbol{\theta}}]_2 + y_1 \quad [\hat{\boldsymbol{\theta}}]_3 + z_1]^T. \tag{2.34}$$

In 2D, $L = 3$ produces ambiguity and the solution depends on the sensor position. In 3D, it is necessary to have a system with minimum four receivers, $L = 4$ in Equation (2.24) to perform a position estimation in 3D, i. Dividing Equation (2.24) by $r_{TDOA,l}$, we have:

$$\begin{aligned} &\frac{2(x_1 - x_l)(x - x_1)}{r_{TDOA,l}} + \frac{2(y_1 - y_l)(y - y_1)}{r_{TDOA,l}} - 2R_1 + \frac{m_{TDoA,l}}{r_{TDoA}} \\ &= r_{TDoA,l} - \frac{(x_1 - x_l)}{r_{TDoA,l}} - \frac{(y_1 - y_l)}{r_{TDoA,l}}, \qquad l = 2, 3, \ldots L. \end{aligned} \tag{2.35}$$

The LLS is a straightforward implementation, but it only shows better performance if the noise is independent and identically distributed. As described in [Rao et al., 1973], it is possible to improve the accuracy if we include symmetric weighting matrix $W$ in the cost function.

The WLLS version for Equation (2.21) denoted by $J_{WLLS,TDoA}(\tilde{\boldsymbol{x}})$, would be:

$$\begin{aligned}\mathbf{W} &= [4diag(d_2, d_3, \ldots d_L)\mathbf{C}_{TDOA}diag(d_2, d_3, \ldots d_L)]^{-1} \\ &\approx [4diag(r_{TDOA,2} + R_1, r_{TDOA,3} + R_1, \ldots r_{TDOA,L+R_1})]\mathbf{C}_{TDOA} \\ & diag((r_{TDOA,2} + R_1, r_{TDOA,3} + R_1, \ldots r_{TDOA,L+R_1}))^{-1}\end{aligned} \tag{2.36}$$

$$\begin{aligned}\mathbf{W} \approx \big[&4diag(r_{TDOA,2} + [\hat{\theta}]_3, r_{TDOA,3} + [\hat{\theta}]_3, \ldots, r_{TDOA,L} + [\hat{\theta}]_3)\mathbf{C}_{TDOA} \\ &diag(r_{TDOA,2} + [\hat{\theta}]_3, r_{TDOA,3} + [\hat{\theta}]_3, \ldots, r_{TDOA,L} + [\hat{\theta}])\big]^{-1}\end{aligned} \tag{2.37}$$

Matrices $\mathbf{A}, \boldsymbol{\theta}, \mathbf{b}$ are defined in Equations(2.25), (2.26) and (2.27), respectively. It is necessary to estimate the value for $R1$ using the LLS.

### 2.3.3. Comparison of Localization Algorithms

Table 2.1 summarizes the key aspects to compare the different algorithm implementations for TDoA localization, based on [Zekavat and Buehrer, 2011].

Usually, nonlinear approaches are more complex, require more computational effort but present higher accuracy. They may not converge to a feasible result because the better solution search procedure can end up in a local maximum instead of the global one.

The linear approach tends to be more straightforward and also gives good accuracy. It is necessary to have the noise statistics to perform the correct linearization in the position estimation. The knowledge of noise statistics is not part of the framework of the method proposed.

We have implemented the NLS, ML, as nonlinear approach; and the LS and WLS, as

Table 2.1.: Classical Algorithm for TDoA, adapted from [Zekavat and Buehrer, 2011].

| Algorithm | Positive Aspects | Negative Aspects |
|---|---|---|
| NLS | - Achieved accuracy is high.<br>- Do not need the noise statistics. | - Global solution may not be guaranteed.<br>- Using grid or random search high complexity. |
| ML | - Accuracy is high. | - Global solution may not be guaranteed.<br>-Complexity is high, if a grid or random search is involved.<br>- Noise statistics are needed. |
| LLS | - Global solution is guaranteed.<br>- Simple and computationally efficient.<br>- Noise statistics are not needed. | - Low accuracy. |
| WLLS | - Global solution is guaranteed.<br>- Highest accuracy can be achieved with constraints. | - Noise statistics are needed.<br>- many iterations may be required. |

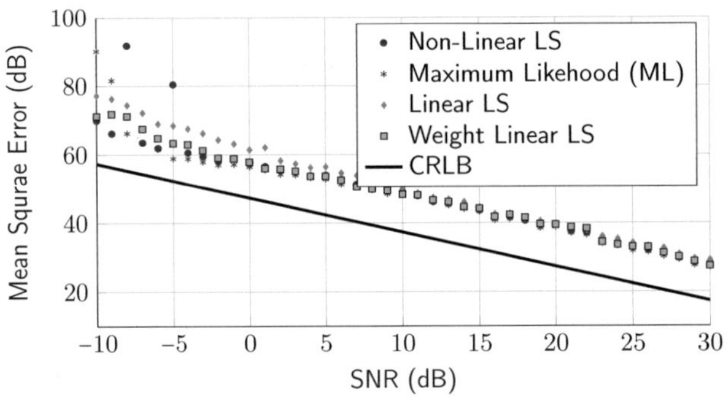


Figure 2.2.: Comparison TDoA Algorithm.

linear one. The Figure 2.2 shows the performance of the methods in a simulated scenario assuming $L-1$ distinct time-differences. The base sensor is considered the reference; we have added noise as a Gaussian-zero mean process.

The comparison was made, analyzing the mean square error, variating the Signal to Noise Ratio (SNR) within the range [-10,30] dB. We included also the classical CRLB derived in [Torrieri, 1984]. As indicated in Figure 2.2, with lower SNR, NLS and ML have better performance, whereas in large SNR there are almost no difference. All of them converged asymptotically to the CRLB values.

We should note, for lower SNR values, the error of the nonlinear approach is higher, showing outliers caused by the non-convergence situation or the existence of a local maximum in the optimization search for the solution.

## 2.4. Practical Aspects in TDoA Systems

This section analyses the impacts of miss-synchronization, multipath, bandwidth, and sample frequency constraints on the system's performance. In the most simple approximation, a TDoA sensor [Kolb et al., 2012] is a wideband receiver, with some processing means to sample, synchronize, and transmit a set of recordings to a central station so that it may perform the position estimation.

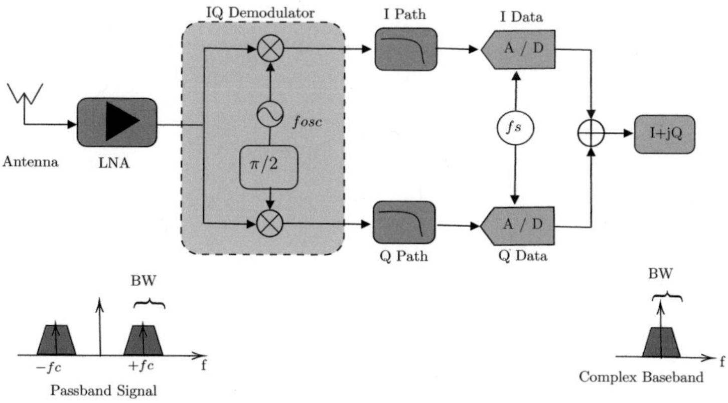


Figure 2.3.: Basic TDoA Sensor Architecture.

The Figure 2.3 shows the basic structure of a TDoA sensor, where the signal received is filtered, down-converted, and sampled, during a "time-window" (integration time $T_i$), where each complex number (I+jQ) corresponds a Complex Baseband (CBB) sample. Recapturing once more that we are dealing with a non-collaborative outdoor situation, there is no information either collaboration among the sensors and the target emitter. The critical aspect in the system is the synchronization among the nodes and the desired spatial resolution, as pointed out in [Kolb et al., 2012].

Figure 2.4 and Figure 2.5 show the complete framework of a functional TDoA localization system, including sensor nodes recording, sampling, and transmitting samples of the desired signal to the central station to be processed. It can be seen that an actual outdoor system has several constraints to be deployed.

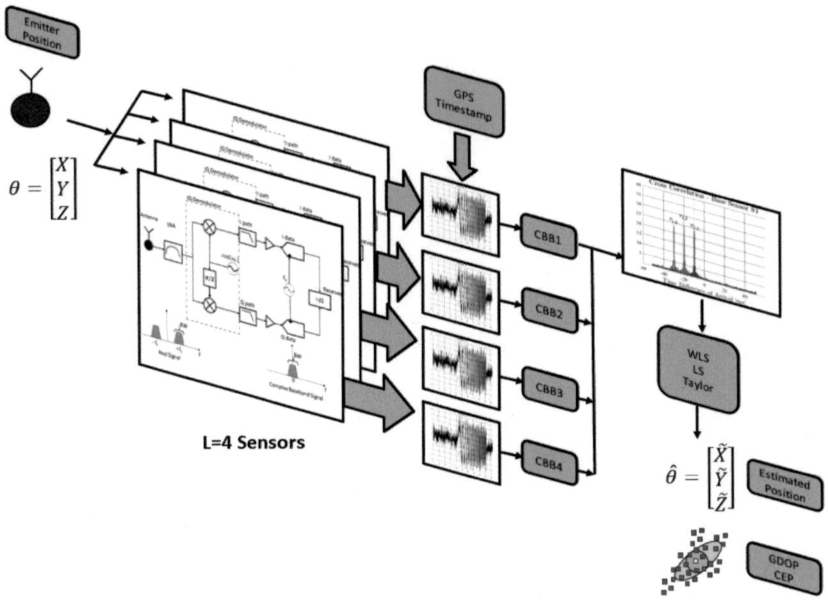


Figure 2.4.: TDoA System Framework.

In non-collaborative localization, the system does not have information about the target to be located. Hence, the sampled frequency, the recording time, how to deploy the system, and several other aspects need to be set to improve the system performance.

The central node, or the fusion center, performs the position estimation using the files received from the remote nodes recursively. There are several tracking approaches to keep the position and to follow the target fusing the results to enhance precision [Degen et al., 2014].

A precise estimation requires that the signal in each sensor be down-converted, filtered, buffered, and sent to the fusion center to be processed. Each of these steps is a system design decision that impacts the sensor bandwidth, the sampled frequency, the SNR, and the strategies to cope with multipath effects.

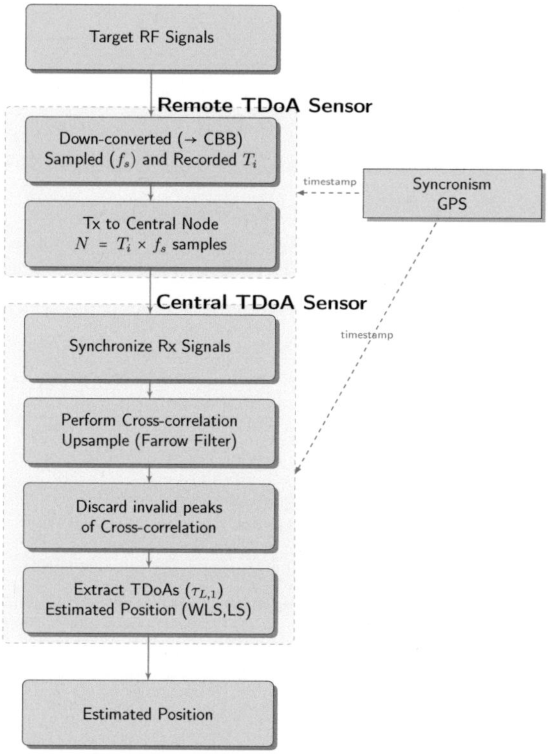


Figure 2.5.: Practical Aspects of TDoA System.

### 2.4.1. Synchronization Requirements in TDoA

TDoA systems usually present strict synchronization requirements as the principal drawback; most signal processing techniques use a two-step approach. First, the sensors extract the received parameters; then, the emitter position is estimated in a second step. Precise synchronization between receivers is essential for achieving high accuracy. The Figure 2.6 shows two conventional approaches to synchronize sensor nodes, either using Global Positioning System (GPS) timestamps or a transmitter in a known position as a reference.

In such a design, receivers must be precisely time-synchronized, which adds computational costs and brings errors that may decrease localization accuracy [Xu et al., 2012]. Although a non-collaborative localization system does not require timestamps of the target signal, a time reference among the sensors is necessary to extract the correct time differences.

The nonlinear relationship between TDOA measurements and position makes the synchronism errors play a significant role in position estimation variation. Due to the high propagation speed of electromagnetic waves, the most critical challenge is how to insert a high-precision timestamp in each sensor.

In the GPS (also GLONASS or Beidu), the timing reference accuracy depends on signal reception quality. The GPS is also affected by ionosphere burst, multipath, and imperfections in the clock circuit stability in each node [Cho et al., 2012b]. A significant challenge is time offsets mitigation between the clocks used in each sensor. It is possible to model the deviation as an additional variable in the system, where inaccurate and unstable circuit clocks increase the overall error.

There are clock synchronization mechanisms for eliminating offsets and drifts, but they always remain with a certain error level. Multipath and ionospheric refraction problems are inherent of the outdoor channel propagation; they can be statistically modeled but not eliminated [Byun et al., 2002].

Extensive measurements campaign to analyze GPS offset drift errors in TDoA systems was carried out by [Li et al., 2013]. After 12 hours of quantitative evaluations, they showed an average of $200\ ns$ clock offset between the two sensors, giving a positioning error of 60 meters. In normal conditions, the synchronization error is lower than $30\ ns$ Root Mean Square (RMS), which corresponds to an accuracy of 9 meters.

It is essential to use a suitable synchronization method to obtain absolute timestamps independent of the receiver settings. Typically, the position of the sensor is arbitrarily in the scenario, which results in receiver position errors. Several investigations [Ho et al., 2007], and [Kovavisaruch and Ho, 2005] showed that both receiver position errors and synchronization clock bias in TDOA localization degrade the positioning accuracy.

In our studies, the TDoA system uses the Global Position System as a timestamp provider, assuming that all sensors can track the same GPS set. The position processing uses the recorded GPS signal with the CBB to estimate the emitter position. Based on [Li et al., 2014], the synchronism influences on TDoA can be modeled by Equation (2.38).

$$\tau_{TDoA} = T_d + T_{mp} + T_{sy} + T_n, \tag{2.38}$$

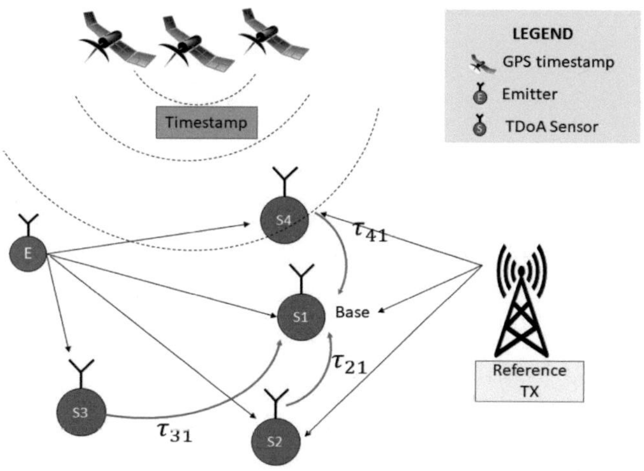


Figure 2.6.: TDoA synchronization with GPS timestamp or Reference Emitter.

where $\tau_{TDoA}$ is the time-difference estimated. The $T_d$ is the actual range difference taking into account the sensor and the emitter geographic position divided by the speed of ligth $c$. The term $T_{mp}$ is the extra-time produced by a reflection or diffraction from the multipath channel effects. $T_{sy}$ is the error in the synchronization. The last item, $T_n$ corresponds to Gaussian noise modelling other possible error sources.

If we restrict our analysis for the static environment with a fixed position of the target, $T_d$, $T_{mp}$ and $T_{sys}$ would be constant, while the noise is Gaussian. There are several approaches to deal with synchronization errors for increasing the performance of a TDoA [Aljadeff et al., 2010] and [Behrendt et al., 2006]. This synchronization issue is out of our scope, and we will concentrate on the multipath error part, defined here as a constant bias on the TDoA measurements.

### 2.4.2. Accuracy in TDOA Measurements

We have started this chapter formulating the localization processing based on TDoA as a geometric problem, with Gaussian zero-mean noise. Hence, there are several issues to be defined to extract time differences from the receiving signal correctly. All of the methods of computing the geolocation are subject to errors in the measurements. The accuracy of TDOA measurements is discussed in this section in terms of the standard deviation.

It is not the scope of our research to design a complete TDoA system. However, it is necessary to explain the actual constraints of such systems to understand how the proposed method can enhance the inherent limitation of outdoor applications. We start by mentioning that measurement errors result in wider TDoA hyperbolas, increasing the regions within which the target should lie, as illustrated in Figure 2.7.

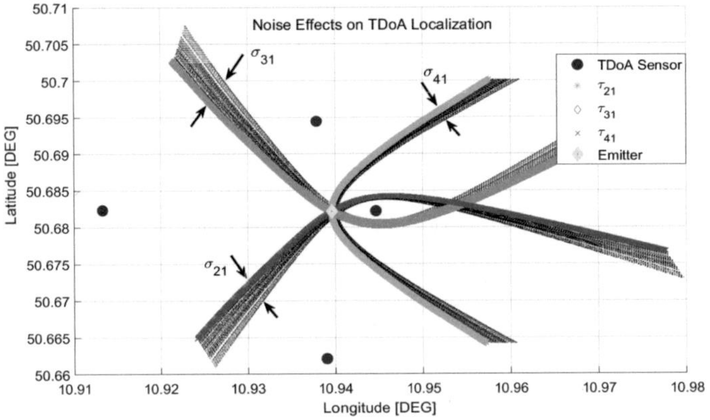


Figure 2.7.: TDOA isochrones with errors included.

There is a trade-off between how much bandwidth the receiver should record and the ideal sampling frequency. One of the possible visualizations of the error in the localization processing is the isochrones. An isochrone is defined as a line on a map or diagram connecting places from which it takes the same time to travel to a certain point, [Dictionary, 1989].

The more the noise and the imperfections are included in the calculation, the wider the isochrones become. The hyperbolic relationship makes the distribution of the sensors also affect the result, as the inner area presents a lower error than the asymptotic regions in the peripheries.

Assuming that $S(f)$ is the spectrum of the target signal, the RMS measured bandwidth ($\beta_{RMS}$) can be obtained by [Poisel, 2008]:

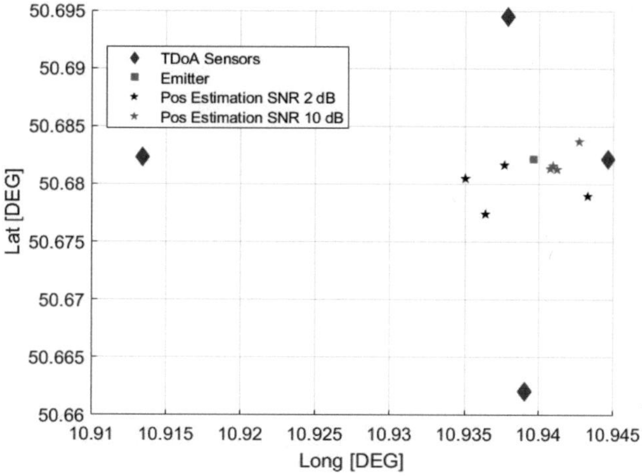


Figure 2.8.: Effects of $SNR$ of TDOA errors.

$$\beta_{RMS} = \sqrt{\frac{\int_{-\infty}^{\infty} |S(f)|^2 f^2 df}{\int_{-\infty}^{\infty} |S(f)|^2 df}}, \tag{2.39}$$

If the signal spectrum is given by $W$ and has sharp edges, the bandwidth can be approximated by $\beta_{RMS} = \pi W/\sqrt{3}$ Hertz.

The signal received in a sensor pair $i, j$ presentes a SNR given by $\gamma_i$ and $\gamma_j$,in each sensor. The joint SNR from the two sensors is $\xi_{eff}$, and is approximated by [Poisel, 2008]:

$$\frac{1}{\xi_{eff}} = \frac{1}{2}\left[\frac{1}{\gamma_i} + \frac{1}{\gamma_j} + \frac{1}{\gamma_i \gamma_j}\right]. \tag{2.40}$$

The integration time $T_i$ is defined as the time period included in the correlation processing. The TDoA sensor bandwidth from each CBB is defined as $B$.

The standard deviation of the error in TDoA systems is lower bounded using Equation (2.41) and it depends on noise received bandwidth, time-window employed (also defined as integration time) and SNR of each signal, [Poisel, 2008].

$$\sigma_{TDoA} \geq \frac{1}{8\pi\beta_{RMS} B T \xi_{eff}}. \tag{2.41}$$

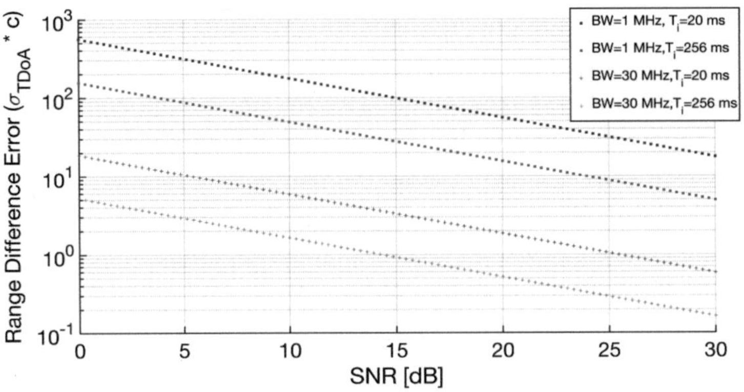


Figure 2.9.: Effect of SNR, $T_i$ and BW on TDoA errors.

Broader bandwidth makes the peak in the cross-correlation function be sharper [Knapp and Carter, 1976]. Measurement errors depend both on the sensor bandwidth and the effective SNR.

The Figure 2.8 shows the effect of SNR in the localization performance. We have used the WLS algorithm and $\xi_{eff} = 2dB$ and $\xi_{eff} = 10dB$. We can notice a better concentration of the estimation when the signal has larger SNR.

In Figure 2.9, we show the TDoA standard deviation using $BW$ of 1 MHz and 30 MHz. In both cases, we have assumed the integration time $T_i$ of 20 ms and 256 ms. We can realize that higher bandwidth causes a higher reduction in errors than the integration time.

### 2.4.3. Signal Processing for TDoA Extraction

The base TDoA sensor is defined as a "central node," and it receives the sampled signal CBB for all the other sensors and performs the cross-correlation to extract the time differences. As discussed by [Poisel, 2008] and [Stein, 1981], one conventional method of determining the time delay between the signals arriving in sensors is to compute to find the argument that maximizes the cross-correlation function of the received signals. The estimated TDoA is the argument that defines a peak in the pairwise cross-correlation function between each pairwise sensor signal.

Considering $s(t)$ represents the emitted signal from the radio frequency target and considering the channel effects including noise [Knapp and Carter, 1976], the received signals arriving in two TDoA sensor, $x_1(t)$ and $x_2(t)$, may be given by:

$$x_1(t) = s(t) + n_1(t),$$
$$x_2(t) = As(t - \tau_{21}) + n_2(t), \tag{2.42}$$

where, $n_1(t), n_2(t)$ are noises, $\tau_{21}$ is theTDoA and $A$ is the amplitude ratio for scaling the signal. $s(t)$ and $n_1(t), n_2(t)$ are real and $s(t)$ is assumed to be uncorrelated with the noise signals. The cross-correlation of this two received signal is given by:

$$R_{x1,x2}(\tau_{21}) = E[x_1(t)x_2(t - \tau_{21})], \tag{2.43}$$

where E denotes expectation.

Because of the finite observation time, however, $R_{x1,x2}$ can only be estimated.Assuming a finite observation time, the estimated cross-correlation $\hat{R}_{x1,x2}(\tau)$ is:

$$\hat{R}_{x1,x2}(\tau) = \frac{1}{T - \tau} \int_{\tau}^{T_i} x_1(t)x_2(t - \tau)dt. \tag{2.44}$$

The term $T_i$ represents the observation interval. The time delay where the cross-correlation has a maximum is the estimated TDoA, $\tau_{21}$. Figure 2.10 shows a simulation of an BPSK-OFDM signal, with SNR of 10 dB, $N = 64$ sub-carriers($Ns_{data} = 48$ data sub-carries and $Ns_{pilots} = 4$ pilot sub-carries), $BW = 20MHz$, period=$3.2\mu s$.

One difficulty that arises from this is that data sharing requires significant bandwidth. Specifically, the cross-correlation of two received signals requires that one of the two sensors transmit an extended portion of the received signal to the other sensor included in the computation. However, this may require a large amount of bandwidth and power, which are limited resources for mobile, wireless devices.

It is relevant to observe that a CBB signal with $20\ MHz$ sampled at $25\ MHz$ during $20\ ms$ produces 500,000 complex samples to be transmitted to the central node for each sensor. The sampling frequency used defines the spatial resolution achieved, which is a constraint in the system deployment.

As the scope of this research is a "non-collaborative" localization system, the sensor

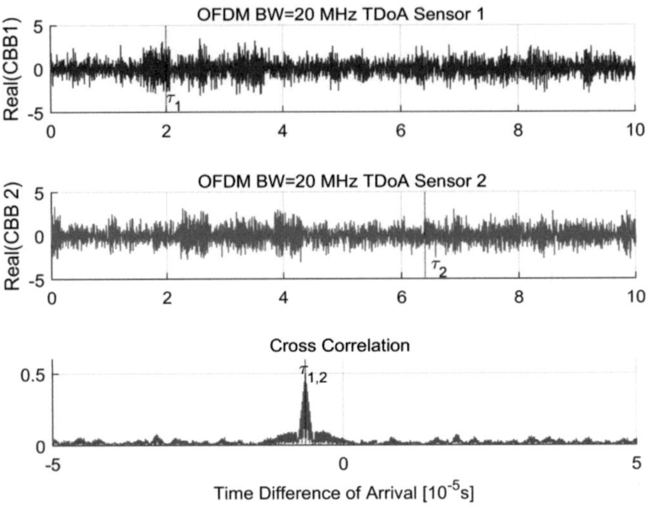


Figure 2.10.: TDOA extraction OFDM signal with BW 20MHz.

has no information about the target signal. Thus, there is a trade-off between how much bandwidth the receiver should record and the ideal sampling frequency.

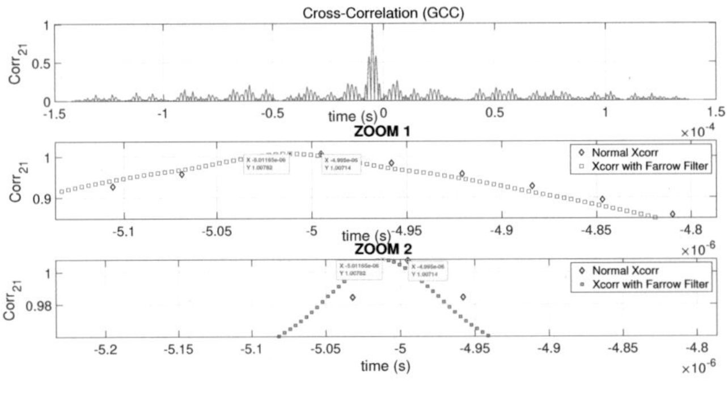


Figure 2.11.: Time domain extraction using Farrow Filter for Oversampling.

It is possible to apply the oversampling procedure to improve the TDoA spatial resolution mathematically. One approach is to interpolate the samples and smooth the cross-correlation function, enhancing the estimation and defining the maximum point better.

The interpolation of the time dependant signals at the receivers can be done using

Farrow filters [Kolb et al., 2012], and the continuous smooth curve may be parabolic. Figure 2.11 shows the effect of cross-correlation oversampling resolution. The red graphic is an interpolated version of the "blue" one, where we can achieve a more accurate TDoA.

### 2.4.4. Quality and Power Measurements in TDOA Estimation

Channel imperfections, including noise and multipath, affect TDoA estimations. As described by [Kolb et al., 2012], after the calculation of the cross-correlation function, two parameters are calculated to evaluate if the measurements should be taken or discharged in the position estimation processing: the Cross Power Spectral Density (CPSD) and the Quality.

Assuming the CBB signal in two TDoA sensors are $y(t)$ and $x(t)$, their power spectral densities, $S_{yy}(\omega)$ and $S_{xx}(\omega)$ are the Fourier transform of their auto-correlation functions, as follow:

$$S_{xx}(\omega) = F\left\{R_{xx}(\tau)\right\}, \quad S_{yy}(\omega) = F\left\{R_{yy}(\tau)\right\}. \tag{2.45}$$

The estimated power in $\hat{R}_{x1,x2}(\tau)$ is calculated using the observation time $T_i$ of the time signal data segment from each CBB. It is usual to represent this value in $dBm$, the length of the data segment is defined by the *integration time* and the sampled frequency used.

As described in [Aatique, 1997], the correlation function is estimated over "Window" defined by observation interval $T_i$. This parameter affects the precision in TDoA estimation, increasing the correlation window size, the auto-correlation of a signal shows a more narrow peak [Montminy, 2007].

The CPSD is the Fourier Transform of the pairwise cross-correlation function between sensors [Ahmed et al., 2013] and defined as:

$$\hat{S}_{xy}(\omega) = F\left\{\hat{R}_{xy}(\omega)\right\} = \int_{-\infty}^{\infty} R_{xy}(t)e^{-j\omega t}dt.. \tag{2.46}$$

The CPSD can be obtained with Fourier transform of the cross-correlation function,

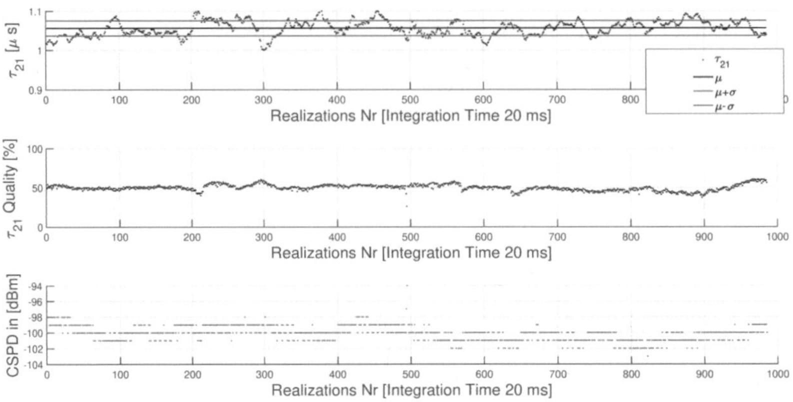


Figure 2.12.: Multipath extra-error produced in TDOA Location.

as follows:

$$S_{xy}(\omega) = \int_{-\infty}^{\infty} R_{xy}(t)e^{-j\omega t}dt = \int_{-\infty}^{\infty}\left[\int_{-\infty}^{\infty} x(\tau)y(\tau+t)d\tau\right]e^{-j\omega t}dt, \tag{2.47}$$

where the value $R_{xy}(t)$ represents the cross-correlation of $x(t)$ and $y(t)$.

Assuming that the $CBB$ signals are discretized and sampled by $f_s$, giving $x_n$ and $y_n$; the CPSD and the cross-correlation are related by:

$$S_{xy}(\omega) = \sum_{n=-\infty}^{\infty} R_{xy}(n)e^{-j\omega n}, \tag{2.48}$$

where

$$R_{xy}(n) = E\left[x_n y_{n-\tau}^*\right]. \tag{2.49}$$

The second criteria is the **Quality** in the TDoA estimation. For each realization of cross-correlation, the peak in the estimated cross-correlation is normalized and compared and normalized from 0 to 1 (0% - 100%). This parameter is related to the variance in the TDoA measurements; a quality factor is assigned to the highest peak in the correlation function. It is usual to perform position estimation with delays with **Quality** greater than $50\%$.

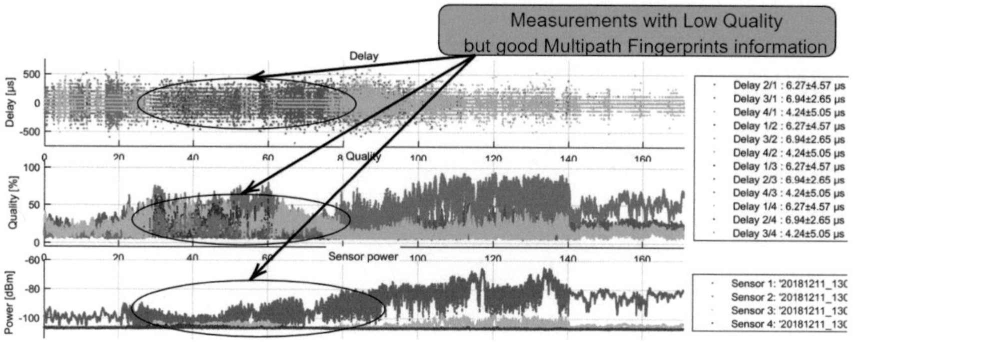


Figure 2.13.: Effects of Multipath in TDoA results.

There is a dependency of this parameter with the multipath contribution. If we decide to filter out the lower values, also the multipath contributions will be extracted out from the measurements [Akula et al., 2016].

Figure 2.12 show the multipath effect in a in only one TDoA ($\tau_{21}$) estimation, while Figure 2.13 shows the quality and CPSD for the enhanced set of all TDoA using 4 sensors. As we can observe, there are variations in the TDoAs estimations. Looking at the Quality and the Power graphics, we notice that the strong variations occur when the cross-correlation either has low power or bad qualities. It is a key aspect in our formulation because we will try to extract the multipath information using the quality and CPSD as an outlier filter, as described by [Kolb et al., 2012].

### 2.4.5. Effect of NLOS in TDOA Estimation

This section has formulated the data model to support the difficulty of estimating a radio-frequency emitter in a NLOS scenario. Suppose the TDoA system is deployed in a dense multipath scenario. In that case, there are multiple peaks in the correlation function, resulting from the sum of delayed versions of the received signal.

In a wide-sense stationary channel [Falk et al., 2003], the multipath produces a bias in the estimations as indicated in Figure 2.14, causing an extra noise in the estimation performance.

An overlapping among the peaks of two maximums in the correlation function can

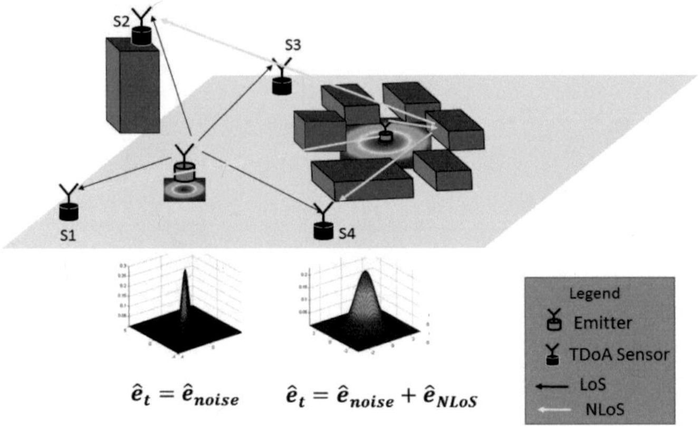


Figure 2.14.: Multipath extra-error produced in TDOA Location.

cause wrong delay estimations. The two rays can add up in the correlation and result in the broader peak, which gives incorrect delay information.

It is possible to use advanced signal processing techniques like compress sensing [Salari et al., 2018], Singular Value Decomposition (SVD), or ML to precisely estimate the time delay in such cases, even with narrow bandwidth signals. We have not analyzed these approaches because we are focusing on describing the TDoA multipath as a localization constraint.

In TDoA systems it is required that the remote stations send to the central sensor a sample of the recorded signal, to extract the time-differences. However, this may require a large amount of bandwidth and power, which are limited resources for wireless devices. Revising once more Equation (2.7), we include a "bias" term $\boldsymbol{b}$ and the system of equations becomes:

$$\boldsymbol{r}_m = \mathbf{f}_m(\boldsymbol{x}) + \boldsymbol{b}_m + \mathbf{n}_m, \qquad m = 1, 2, \ldots, M; \tag{2.50}$$

where $b_i$ represents the measurement $i^{th}$-bias $i$ and $\boldsymbol{x} = \begin{bmatrix} x_T & y_T & z_T \end{bmatrix}^{\mathsf{T}}$ is the target location.

As $\mathbf{f}_m(\boldsymbol{x})$ is nonlinear [Poisel, 2008], we can use a Taylor expansion to find the linear terms that, in matrix form, become:

$$\boldsymbol{r} = \mathbf{f}(\boldsymbol{x}) + \boldsymbol{b} + \mathbf{n}. \tag{2.51}$$

The LS estimator tries to find $\hat{\boldsymbol{x}}$ to minimize:

$$Q(\boldsymbol{x}) = [\boldsymbol{r} - \mathbf{f}(\boldsymbol{x}) - \boldsymbol{b}]^T \mathbf{C}_{\Delta r \Delta r}^{-1} [\boldsymbol{r} - \mathbf{f}(\boldsymbol{x}) - \boldsymbol{b}]. \tag{2.52}$$

If the noise elements $(\boldsymbol{n})$ are zero-mean Gaussian, the noise covariance matrix is $\mathbf{C}_{nn}$. We assume the bias $\boldsymbol{b}$ is unknown with nonrandom parameters. The NLOS characteristics stay constant during the interval between measurements, letting us focus on the most significant multipath contributions. It is also assumed that the target remained fixed over the measurement interval. The Taylor expansion around reference point $(\boldsymbol{x}_0)$ is:

$$\mathbf{f}(\boldsymbol{x}) \approx \mathbf{f}(\boldsymbol{x}_0) + \mathbf{G}_0(\boldsymbol{x} - \boldsymbol{x}_0), \tag{2.53}$$

with,

$$\mathbf{G}_0 = \begin{bmatrix} \frac{\partial f_1(\boldsymbol{x})}{\partial x_1}\Big|_{\boldsymbol{x}=\boldsymbol{x}_0} & \cdots & \frac{\partial f_1(\boldsymbol{x})}{\partial x_1}\Big|_{\boldsymbol{x}=\boldsymbol{x}_k} \\ \vdots & \ddots & \vdots \\ \frac{\partial f_M(\boldsymbol{x})}{\partial x_1}\Big|_{\boldsymbol{x}=\boldsymbol{x}_0} & \cdots & \frac{\partial f_M(\boldsymbol{x})}{\partial x_1}\Big|_{\boldsymbol{x}=\boldsymbol{x}_k} \end{bmatrix}, \tag{2.54}$$

using Equation (2.50) into Equation (2.51),

$$Q(\boldsymbol{x}) = [\boldsymbol{r}_1 - \mathbf{G}_0\boldsymbol{x} - \boldsymbol{b}]^T \mathbf{C}_{\Delta r \Delta r}^{-1} [\boldsymbol{r}_1 - \mathbf{G}_0\boldsymbol{x} - \boldsymbol{b}]. \tag{2.55}$$

To find the LS solution, we need to minimize the function in Equation (2.55), where:

$$\boldsymbol{r}_1 = \boldsymbol{r} - \mathbf{f}(\boldsymbol{x}_0) + \mathbf{G}_0\boldsymbol{x}_0. \tag{2.56}$$

To find the minimum, we need to extract the value that leads to the minimum in the gradient vector,

$$\nabla_{\boldsymbol{x}} Q(\boldsymbol{x}) = \begin{bmatrix} \frac{\partial Q}{\partial x_1} & \frac{\partial Q}{\partial x_2} & \cdots & \frac{\partial Q}{\partial x_k} \end{bmatrix}^T, \tag{2.57}$$

i.e., the value of $\boldsymbol{x}$ that makes $\nabla_{\boldsymbol{x}} Q(\boldsymbol{x}) = \mathbf{0}$. If the bias vector $\boldsymbol{b}$ is assumed not to be a function of $\boldsymbol{x}$ in some proximity of the sensor, giving $\nabla_{\boldsymbol{x}}\boldsymbol{b} = \mathbf{0}$ and $\mathbf{G}_0^T \mathbf{C}_{\Delta r \Delta r}^{-1} \mathbf{G}_0$, then the value of $\boldsymbol{x}$ that gives the minimum $Q(\boldsymbol{x})$ is obtained by:

$$\hat{\boldsymbol{x}} = (\mathbf{G}_0^T\mathbf{C}_{\Delta r\Delta r}^{-1}\mathbf{G}_0)^{-1}\mathbf{G}_0^T\mathbf{C}_{\Delta r\Delta r}^{-1}\boldsymbol{r}_1 - (\mathbf{G}_0^T\mathbf{C}_{\Delta r\Delta r}^{-1}\mathbf{G}_0)^{-1}\mathbf{G}_0^T\mathbf{C}_{\Delta r\Delta r}^{-1}\boldsymbol{b}. \tag{2.58}$$

If the biases are known, meaning that we have some constant in the multipath, the value is stable during the TDoA measurements interval, and Equation (2.58) can be used to determine the best estimate of the target location. If the biases are not known, we can assume that it is zero [Poisel, 2008], and we can use Equation (2.56), as follows:

$$\hat{\boldsymbol{x}} = (\mathbf{G}_0^T\mathbf{C}_{\Delta r\Delta r}^{-1}\mathbf{G}_0)^{-1}\mathbf{G}_0^T\mathbf{C}_{\Delta r\Delta r}^{-1}[\boldsymbol{r} - \mathbf{f}(\boldsymbol{x}_0)] + \boldsymbol{x}_0. \tag{2.59}$$

and with Equation (2.53), we have:

$$\hat{\boldsymbol{x}} = (\mathbf{G}_0^T\mathbf{C}_{\Delta r\Delta r}^{-1}\mathbf{G}_0)^{-1}\mathbf{G}_0^T\mathbf{C}_{\Delta r\Delta r}^{-1}[\mathbf{f}(\boldsymbol{x}) - \mathbf{f}(\boldsymbol{x}_0) - \mathbf{G}_0(\boldsymbol{x} - \boldsymbol{x}_0) + \boldsymbol{b} + \boldsymbol{n}] + \boldsymbol{x}. \tag{2.60}$$

We have performed a simulation of a QPSK-OFDM signal with CBB having $BW = 2MHz$, in the City of Ilmenau, Germany, using TDoA sensors deployed with the geometry illustrated in Figure 2.15. It indicates how the bias affects the TDoA system, where we have used the Confidence Ellipsis with 99% ($GDOP_{99}$). The noise was assumed as Gaussian, and the bias in the measurements $r_{31} = \tau_{31}c = 100\ m$ and $r_{21} = \tau_{21}c = 80\ m$, following [Koorapaty et al., 1998] and [Poisel, 2008].

The blue circles indicate the TDoA sensors. The $GDOP_{99}$ calculated with 100 estimations shows the error lower than 120 meters, with some outliers more than 300 m. The ellipsis center, representing the mean values, was not the target position, indicating that the bias changed the estimations according to the multipath effect.

## 2.5. Concluding Remarks

The analysis presented in this chapter gave a geometric and mathematical formulation, describing the relationship between the range distances and the position. We have explained and simulated some standard algorithms used in TDoA, showing the nonlinear equations to solve the position estimation problem.

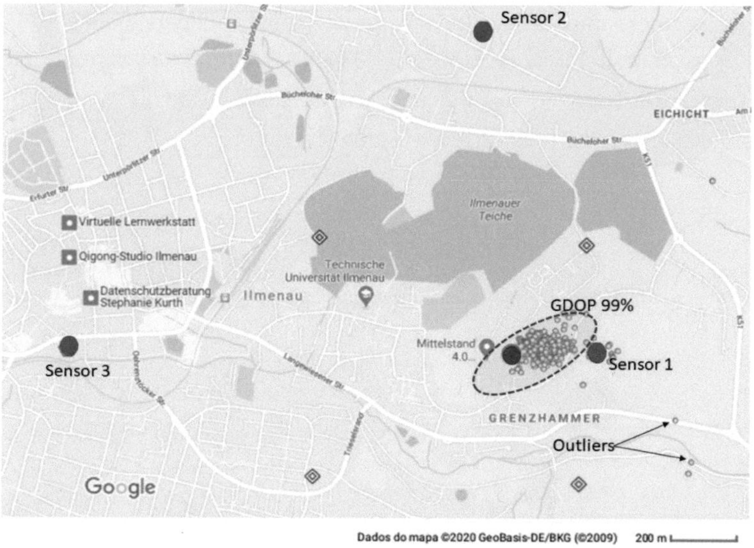


Figure 2.15.: Effects of Multipath modeled as bias $\boldsymbol{b}$ in TDoA results.

We have analyzed the effects of bandwidth, $SNR$, synchronism requirements, and integration time on system precision. We also have derived and explained the localization performance tools to compare different position estimation approaches, including CEP, GDOP,CRLB and Cumulative Distribution Function (CDF).

We also simulated the effects of oversampling in the final TDoA estimations regarding the signal processing issues. The most significant limitation in TDoA is the high capacity communication link requirements to connect the nodes. It is necessary to correlate arrival times between sensors, requiring exchanging large amounts of data as a requirement for real-time operation.

All concepts introduced here are building blocks to understand the proposed method; they are essential to describe the framework in Figure 2.6. It is also worth emphasizing the derivation of Section 2.4.5.

A key point shown in this chapter is that the multipath effects in the cross-correlation function cause extra peaks, which can be removed from the position estimation processing. Additionally, it is also relevant in localization methods that if the NLOS situation is a wide-sense stationary channel, the bias remains the same during the integration window.

Finally, and most importantly, this chapter aimed at understanding the effects of multipath in the overall TDoA performance. This NLOS condition corrupts the position estimation in classical algorithms, and our approach is to use this as an extra layer of information, enhancing the final position estimation result.

# 3. RF Channel Estimation for Localization

This chapter reviews concepts and models for wireless channels and their impact on their overall localization systems. In the wireless community, Channel Impulse Response (CIR) refers to a process of extracting the propagation properties of the transmitter-receiver link, including the combined effect of scattering, fading, and power decay with distance.

Time delay-based passive localization relies on the dynamic reflection path that corrupts the required estimation of propagation delays. Moreover, the signal bandwidth is usually not large enough to reach the defined delay resolution. RF localization systems use sensors deployed in a region, each having a CIR that describes the emitter-sensor link.

In a macro sense, TDoA systems are Single Input Multiple Output (SIMO) structures, where the channel characteristics are embedded in the CBB signals produced by each sensor. The framework developed in our research tries to use multipath information to enhance localization processing. Still, the crucial question is how to find a way to extract the delays and the amplitude of the echoes in the received signal accurately.

We start using the general description of the full polarimetric channel model based on [Landmann, 2008] and [Richter, 2005], defining the specular components and the Dense Multi-path Component (DMC).

An outdoor wireless channel model has several parameters to describe the reality adequately; we have simplified our analysis by employing a data model based on a finite number of specular propagation paths. Consequently, we reduce the complete model to consider only delays and amplitude once we work with single-antenna devices.

It is known that this data model is not sufficient to describe all types of radio channel observations. Still, under certain conditions [Vaughan and Andersen, 2003], the Specular Component (SC) do have a more significant influence on the overall system performance.

In the following sections, we analyze the impacts of the limitation in bandwidth and the signal length on the multipath estimation process. To overcome these limitations, we use several realizations for extracting the CIR, introducing a statistical analysis improving the time-delay estimation.

We have studied and implemented two approaches for multipath extraction. In the first one, we have modeled a TDoA system as SIMO and applied a blind Finite Impulsive Response (FIR) equalization scheme using eigenvalue decomposition in the received CBB signals. In the second method, we used the volume cross-correlation [Shi et al., 2016] produced by the TDoA sensor pairs. We showed that the multipath patterns could be associated with the peaks in the Volume Cross-Correlation (VCC) and give positioning feature fingerprint.

Thus, we highlight the limitations of the introduced radio channel model, emphasizing the conditions where specular components have a notable influence in the CIR, like suburban and campus-like conditions. We carried out simulations to evaluate the performance of multipath extraction, considering the impacts of bandwidth, the SNR and the number of realizations.

## 3.1. Motivation and Related Work

This study aimed to develop and validate the localization of radio-frequency emitter in an outdoor scenario in a non-collaborative way. Thus, the collection of related work aims to show several building blocks to understand how to extract the multipath information from the distributed received signals in a blind procedure.

The multipath estimation is an essential part of the channel sounding procedure or equalization using SIMO FIR. In our approach, we suggest using the multipath information from the CBB of each sensor or the TDoA cross-correlation.

The blind channel estimation in communications can be a signal processing operation for recovering multipath, using nuclear norm minimization as described by [Masmoudi et al., 2012].

[Zeng et al., 2010] proposed a modified projection onto convex sets (POCS) using a low complexity algorithm optimization approach that can estimate the delays in the attenuation factors from a mixture signal with several overlapping echos. The simulations showed a better time resolution and robustness to noise compared with the matched filtering. The idea was an improvement of [Fuchs, 1999], who suggested a deconvolution scheme to estimate the number of the paths using the known transmitted signal.

[Kostic et al., 1992] developed a method for channel parameter estimation using a combination of matched filtering and auto-regressive modeling applied to specular multipath channels. [Park et al., 2017] describes a time delay estimation technique using compressive sensing (CS), also given by [Gui et al., 2010] to exploit the sparsity of the multipath channel for a sparse signal reconstruction problem.

[Mayyala et al., 2017] showed a novel subspace-based method for blind identification of multi-channel finite impulse response systems. [Moulines et al., 1995] exploited the orthogonality property between signal and noise subspaces to build a quadratic minimization form in a channel identification context. Moreover, [Masmoudi et al., 2012] presented a new implementation of the maximum likelihood criterion for estimating the time delays in a multipath environment, suggesting a concept of importance sampling (IS) find the global maximum of the compressed likelihood function.

[Bonna, 2017] introduced a framework to estimate the channel multipath without knowledge of the transmitted signal, named Blind Channel Identification (BCI) which used the sparsity in the Single Input Multiple Output (SIMO) structure of the received signals. [Gui et al., 2010] proposed a sparsity method to exploit multipath channels using a greedy algorithm and convex program.

[Phelan et al., 2012] explores a GSM positioning fingerprints method for determining the location using a multipath signature for each possible transmitter location in an area of interest.

[Hinteregger et al., 2017] also showed a Bayesian estimation to specular multipath components applicable on single-input multiple-output radio channels.

[Setlur et al., 2011], [Setlur et al., 2013b], [Setlur et al., 2013a], [Jaffe and Wax, 2014] and [Mittermayer, 2014] also applied multipath fingerprinting to single-site localization, including real data validation. In another initiative, [Leigsnering et al., 2014] added the concept of compressive sensing (CS) on Multipath exploitation to through-the-wall radar.

[Mahler et al., 2015] studied safety-critical scenario for vehicle-to-vehicle communication, focus on the propagation process for locations of the scattering objects. [Naseri and Koivunen, 2016] explained an indoor positioning scheme using multipath delay estimates, modeled in two-dimensional space with arbitrary node configuration.

The papers chosen were focused on blind channel estimation. As previously discussed in Chapter 1, the localization in our multipath fingerprint method is non-cooperative with the target, so there is no synchronism between the transmitter. It is necessary to estimate the specular components to use multipath information. Most of the state-of-the-art research in the field is related to channel sounding or equalization problems. Our contribution in this field was to apply CIR estimation tools to localization problems, where the SC information is used to build the position fingerprints.

In short, the related work here has collected the significant aspects of the multipath estimation in a distributed radio frequency sensor set, either using only the CBB or using VCC. Each different methods for CIR estimations have pros and cons. There are several evaluation parameters for defining the CIR estimation performance as described in [Zekavat and Buehrer, 2011] such as complexity, sensitivity to the noise, a priori information, and the ability to separate closely spaced multipath.

We have selected the methods that could solve closely spaced multipath components that do not depend on the emitter signal knowledge. SIMO and VCC adopted in our investigations are high-resolution algorithms, based on Blind Source Separation (BSS). In the next section, we explain the proposed method with derivation and simulation.

## 3.2. Channel Model

Figure 3.1 shows the schematic representation of the localization channel model system. In localization problems, the propagation mechanisms, including reflection distractions and refractions, are inserted in the channel data model, which has three parts, handled separately throughout the estimation procedure.

The first component, called SC, is caused by a finite number of interactions, resulting from a deterministic and limited number of specular reflections.

The SC has a geometric interpretation, where the emitter's position, the receiver, and

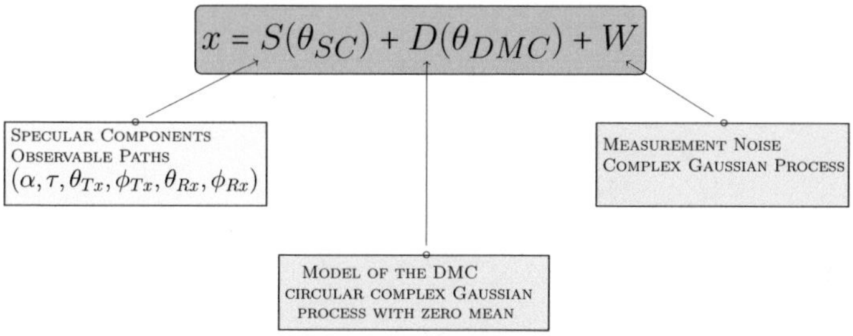


Figure 3.1.: Localization Multipath Channel Model.

the obstacles define the visibility, reflection points, and path description.

The second component, the Dense Multipath Component DMC has a stochastic nature and includes refraction, diffraction, and scattering. The unstructured and complex interaction among the obstacles, transmitters and sensors creates components that contribute to multiple interactions between different reflection and scattering points.

The last component is the noise ($W$), considered a complex zero-mean Gaussian process. The relation between the power transferred by the SC and DMC alters with the scenario type, as described by [Vaughan and Andersen, 2003]. The suburban or Campus-like scenario is often characterized by small and spaced buildings with some open areas in between. In this case, the SC are more defined and transport more energy than DMC components.

Following [Richter, 2005] the channel model can use a complete polarimetric approach, where the Doppler drift, the delay time, the amplitude, and the cross-polarization effect can be included in a unique structure to represent all the channel effects in a compact form.

In the scope of our research, we will concentrate on SC and the geometric interpretation of the path rays. As described by [Landmann, 2008], the total amount of specular propagation paths in a scenario is relatively small. However, they significantly contribute to the total power transferred from the transmit antenna to the receiver.

Following [ITU-R, 2017], the angular delay-Doppler domain is a procedure to define the multipath using a set of appropriate parameters for the statistical description of multipath effects and provides examples of correlation effects among multiple propagation paths and their computation.

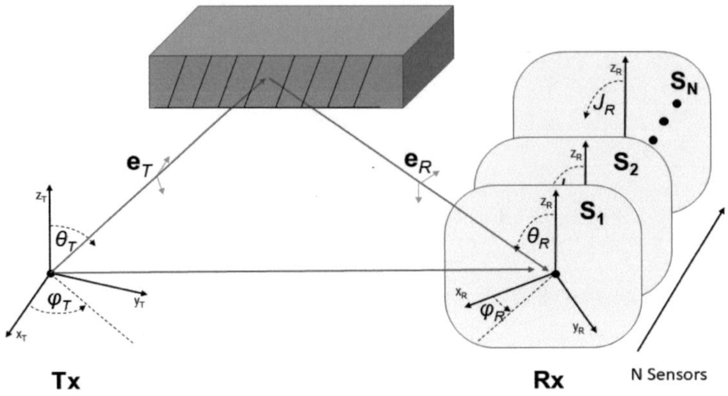


Figure 3.2.: Channel Model for TDoA Localization.

Restricting our analysis to Specular Components [Richter, 2005], each ray connecting the emitter to each of the sensors TDoA can be modeled using the following structural parameters:

- Amplitude($\alpha$): amplitude in V/m of each ray.
- Direction Of Departure (DoD) (Azimuth and Elevation - $\theta_{Tx}, \phi_{Tx}$): The angle from which usually propagating wave leaves the emitter.
- Direction Of Arrival (DoA) (Azimuth and Elevation - $\theta_{Tx}, \phi_{Tx}$): The angle from which usually a propagating wave arrives at TDoA sensors.
- Doppler-shift ($\Delta$): Doppler effect, the apparent difference between the frequency at which radio-frequency waves leave the emitter and that they reach a TDoA sensor, caused by the relative motion of sensor and emitter.
- Delay ($\tau$): delay of each ray.

The discrete angular-delay-Doppler representation of a set of Specular Components is described as a superposition of $K$ functions. This representation uses a complex polarimetric path weight matrix with $2 \times 2$ components $\Gamma_{P1,P2,k}$. $P1, P2$ represents the cross-polarization terms from the transmitter-receiver interaction. The polarization effects can be unified in the matrix form $\mathbf{\Gamma_k}$.

$$\mathbf{\Gamma}_k = \begin{bmatrix} \gamma_{\varphi\varphi,k} & \gamma_{\vartheta\varphi,k} \\ \gamma_{\varphi\vartheta,k} & \gamma_{\vartheta\vartheta,k} \end{bmatrix}. \tag{3.1}$$

$$\begin{aligned} \mathbf{H}_\delta\left(\alpha,\tau,\theta_{Rx},\phi_{Rx},\theta_{Tx},\phi_{Tx}\right) = \sum_{k=1}^{\mathrm{K}} \big( [\mathbf{\Gamma}_k] \times & \quad . \\ \delta\left(\Delta-\Delta_k\right)\times\delta\left(\tau-\tau_k\right)\times & \quad . \\ \delta\left(\phi_{Rx}-\phi_{Rx,k}\right)\times\delta\left(\theta_{Tx}-\theta_{Tx,k}\right)\big) & \end{aligned} \tag{3.2}$$

.

[Richter and Thoma, 2005] proposed a complete and self-contained high-resolution polarimetric CIR. This joint estimation of the parameters includes the SC, the distributed diffuse scatter contribution, and the cross-polarization terms to the CIR.

Since we are working with single element antenna sensors in an outdoor scenario, we shrink the polarimetric CIR model, taking out the Doppler shifts and the polarization mismatching. In this case, the emitter, the sensors, and the environment are instantaneous, meaning that they change slowly with time.

The CIR may assume a simpler model, consisting of a delayed version of a known deterministic signal corrupted by an additive white Gaussian noise. In this case, the channel model considered is given by the general expression for the multipath fading channel, as follows:

$$h(t) = \sum_{i=0}^{L-1} \alpha_i \delta(t-\tau_i), \tag{3.3}$$

where $L$ is the number of rays in each emitter-sensor channel, $\tau_i$ is the delay of the $i^{th}$ path $\alpha_i$ is the amplitude.

In a general case, amplitude $\alpha$ assumes a complex value to consider the phase of each multipath component. In our simple model, we assume that the amplitude is a real number. Whenever necessary, we can include the imaginary part in the phase shift description of the wave equation.

In short, the model of the channel describing the effects between the emitter and the set of TDoA sensors includes only amplitudes and delays of the multipath rays.

## 3.3. Multipath Estimation using CBB Signals from each Sensor

This section performs a study of the time-delay estimator to extract multipath information using CBB signals from each sensor. The aim is to analyze the theoretical limits for time delay estimation under outdoor wireless channel propagation conditions. The channel elements are static time-invariant, and the TDoA sensors have long recordings of CBB.

As studied by [Stoica and Nehorai, 1989], the time resolution of multipath estimation is inversely proportional to the system Bandwidth (BW), and it is also affected by the SNR and the number of realizations $N$, see for instance [Kay, 1993].

The recorded signal in each sensor is split into several pieces, where each of them is used in the estimation procedure, known as estimation realization.

Higher SNR and BW yield a better estimation, but the performance is also affected by the number of realizations performed.

It is necessary to extract and precisely estimate the multipath as an additional information layer to TDoA systems. Thus, the bandwidth and precision desired is a trade-off to be optimized.

The non-collaborative scope gives additional constraints to the estimation process; that is why it is relevant to study and to what extent it is possible to improve accuracy by applying oversampling and increasing the number of realizations.

We make the following assumptions throughout this chapter:

1. We have $N$ sensors, and each sensor does the following two types of measurements:

(a) TDOA measurement between sensors which at the end, leads to $(N-1)$ TDOAs per realization;

(b) multipath measurement where $M$ amplitude and delay measurements are estimated per sensor per realization to find $L$ multipath components;

2. We have $N_{re}$ realizations of the above measurements, where $N_{re} > M$.

3. The noises in all the measurements are Gaussian with zero mean.

(a) The noise variance for the TDOA measurements, when converted to distance, is assumed to be $\sigma_1^2$;

(b) The noise variance for the multipath measurements is assumed to be $\sigma_2^2$

4. The measurements in 3.(a) and 3.(b) above are assumed to be independent.

In Equation (3.3), the channel model considered is based on a general expression for the multipath fading propagation, where the delays are $\tau_0 < \tau_1 < \cdots < \tau_{L-1}$, and $L$ is the number of propagation paths.

If the channel is time-invariant or slowly time variant, the $n_{re}$-th observation of the received signal consists of multiple superimposed delayed and attenuated copies of a the transmitted signal $s(t)$ plus noise as Equation (3.4).

$$y(t;n) = \sum_{i=0}^{L-1} \alpha_i(n)s(t-\tau_i) + v(t;n), \quad n = 1,\ldots,N_{re} \tag{3.4}$$

The TDoA sensor receives each $i^{th}$ path exhibits the same signal delayed ($\tau_i$) and with different fading coefficient ($\alpha_i(n)$). Let assume that the unknown amplitude coefficients are considered deterministic, the Gaussian additive noise is zero-mean ($v(t;n) \sim N(0,\sigma^2)$) and the number of realizations, also called as observation, is given by $N_{re}$.

If we apply the Fourier transformed of Equation (3.4), each receiver signal in the frequency domain is given as:

$$Y(w;n) = \sum_{i=0}^{L-1} \alpha_i(n)S(w)e^{-j(w)\tau_i} + V(w;n), n = 1,\ldots,N_{re}, \tag{3.5}$$

where $w = 2\pi f$, $Y(w;n_{re})$ is the receiver signal in frequency domain, $S(w)$ is the transmitted signal in the frequency domain and $V(w;n_{re})$ is the additive noise.

Following the derivation presented by [Serra et al., 2009], sampling the signal in Equation (3.5) at $w_m = w_0 m$, with $m = 0, 1, \ldots, M-1$ and $w_0 = 2\pi/M$, the $Y[m,n]$ can be described in a vector structure using the frequency domain samples.

The vector build with the received signal in the TDoA sensor, $\mathbf{Y}_n \in C^{M\times 1}$, is shown in Equation (3.6).

$$\mathbf{Y}_n = \mathbf{S}\mathbf{E}_\tau \alpha(n) + \mathbf{V}_n \tag{3.6}$$

In Equation (3.6), the components of the diagonal matrix $\mathbf{S} \in C^{M\times M}$ are the frequency samples of $S(w)$. Matrix $\mathbf{E}_\tau \in C^{M\times L}$ contains the delay vectors with the multipath information, the amplitude of each ray (channel coefficients) are the elements of vector:

$$\alpha(n) = \begin{bmatrix}\alpha_0(n) & \alpha_1(n) & \cdots & \alpha_{L-1}(n)\end{bmatrix}^T \in C^{L\times 1}, \tag{3.7}$$

and vector $\mathbf{V}_n \in C^{M\times 1}$ has the noise samples.

Based on [Stoica and Nehorai, 1989], we can figure out that there is an analogy between the multipath delay model of Equation (3.6) and the model used for DoA estimation. The spectral model [Kay, 1993] for several plane waves impinging in a array of antennas is expressed as:

$$\mathbf{y}(n) = \mathbf{A}(\boldsymbol{\theta})\mathbf{x}(n) + \mathbf{e}(n), \quad n = 1, 2, \ldots, N; \tag{3.8}$$

$\mathbf{y}(n) \in C^{M\times 1}$ is the received signal, $\mathbf{x}(n) \in C^{n\times 1}$ is the vector of complex amplitudes and $N$ is the number of snapshots or realizations of the experiment and $\mathbf{e}(n)$ is a complex Gaussian additive noise.

$\mathbf{A}(\boldsymbol{\theta})$, it is a manifold response of the array, with the structure:

$$\mathbf{A}(\boldsymbol{\theta}) = \begin{bmatrix}\mathbf{a}(\omega_0) & \mathbf{a}(\omega_2) & \cdots & \mathbf{a}(\omega_{L-1})\end{bmatrix}, \text{ and,} \tag{3.9}$$

$$\boldsymbol{\theta} = \begin{bmatrix} \omega_0 & \omega_2 & \cdots & \omega_{L-1} \end{bmatrix}. \tag{3.10}$$

Comparing the structure of Equation (3.6) with Equation (3.8), we can do the following analogies [Serra et al., 2009]:

Table 3.1.: Unknowns in Multipath Estimation

| Name | Description |
|---|---|
| $\alpha_i$ | Amplitude of $i^{th}$ ray |
| $\tau$ | Delay of $i^{th}$ ray |
| $N_{re}$ | Number of observations or realization |
| $L$ | Number of Multipath Rays |
| $M$ | Number of Sampled Frequencies |

$$\begin{aligned} A(\boldsymbol{\theta}) &\equiv \mathbf{S}\mathbf{E}_{\tau} \\ \mathbf{x}(n) &\equiv \boldsymbol{\alpha}(n) \\ \mathbf{e}(n) &\equiv \mathbf{V}_n \\ \boldsymbol{\theta} &\equiv \begin{bmatrix} \tau_0 & \cdots & \tau_{L-1} \end{bmatrix}^T \end{aligned} \tag{3.11}$$

The time resolution achieved in multipath estimation is dependent on the SNR and the BW of the signal recorded. What we need to analyze is the problem of improving the estimation with small bandwidth. We need to evaluate if the recording of long signals can improve the estimation.

In the channel formulation, Equation (3.6), with $N$ observations, it is critical to know how precise we can estimate the rays in a blind and non-collaborative way.

Assuming we make $N$ estimations of a channel connecting emitter-sensor that has $L$ multipath Specular Components. The signal received in the frequency domain has $M$ as the maximum value in the frequency representation of the CBB received.

If we assure that the following constrains hold:

(a) $M > L$;

(b) the noise is Gaussian with zero-mean;

(c) the noise in each ray is uncorrelated with the other paths.

If we perform only one observation and apply the results of the derivations presented by [Serra et al., 2009] and [Stoica and Nehorai, 1989], the CRLB for the multipath estimation can be represented by:

$$CRLB(\boldsymbol{\theta}) = \frac{\sigma^2}{2}\left\{\sum_{n=1}^{N} R\left\{\mathbf{H}^H\mathbf{D_E}^H\mathbf{S}^H\left\{\mathbf{I}-\boldsymbol{\Omega}\left(\boldsymbol{\Omega}^H\boldsymbol{\Omega}\right)^{-1}\boldsymbol{\Omega}^H\right\}\mathbf{S}\mathbf{D_E}\mathbf{H}\right\}\right\}^{-1}, \qquad (3.12)$$

where $\boldsymbol{\Omega} = \mathbf{S}\mathbf{E}_\tau$, $\mathbf{H}$ is a diagonal matrix containing the unknown channel fading coefficients,

$$\mathbf{H} = \begin{bmatrix} \alpha_0(n) & & 0 \\ & \ddots & \\ 0 & & \alpha_{L-1}(n), \end{bmatrix}, \qquad (3.13)$$

$R\{\cdot\}$ denotes the real part operator,

$$\mathbf{D_E} = \begin{bmatrix} \frac{\delta \mathbf{e}_0}{\delta \tau_0} & \cdots & \frac{\delta \mathbf{e}_{\tau_{L-1}}}{\delta \tau_{L-1}} \end{bmatrix} = \boldsymbol{\Phi}\mathbf{E}_\tau, \qquad (3.14)$$

and,

$$[\boldsymbol{\Phi}]_{k,l} = \begin{cases} -j\frac{2\pi(k-1)}{M} & , k = l \\ 0 & , \text{otherwise.} \end{cases} \qquad (3.15)$$

As derived in [Stoica and Nehorai, 1989], [Serra et al., 2009] and [Stoica et al., 1989] the CRLB in a more complete formulation, which holds for multipath estimation of $N$ realization takes the form:

$$CRLB(\boldsymbol{\theta}, N\uparrow) = \frac{\sigma^2}{2N}\left\{R\left[\left\{\mathbf{D_E}^{\mathrm{H}}\mathbf{S}^{\mathrm{H}}\{\mathbf{I}-\boldsymbol{\Omega}(\boldsymbol{\Omega}^H\boldsymbol{\Omega})^{-1}\boldsymbol{\Omega}^{\mathrm{H}}\}\mathbf{S}\mathbf{D_E}\right\}\odot\mathbf{P}^{\mathrm{H}}\right]\right\}^{-1} \qquad (3.16)$$

In the Equation (3.16) $N\uparrow$ represents a large number of realizations, $\odot$ is the element-wise product and the term $\mathbf{P}$ is given by:

$$\mathbf{P} = \frac{1}{N}\sum_{n=1}^{N} \boldsymbol{\alpha}(n)\boldsymbol{\alpha}(n)^{\mathsf{H}}. \tag{3.17}$$

The intention of bringing the multipath estimation derivation was to provide a mathematical foundation to the estimation process that regularly repeats in a constant or slow variant scenario.

Assuming, for instance, that an emitter signal was recorded by a TDoA sensor with 1 MHz bandwidth, sampled at fs=1.25 MHz, 200 ms will produce 250,000 complex samples CBB.

The time resolution ($\Delta t$) is 800 ns, representing two closed rays with the length difference of $\Delta d = \Delta t \times 3.10^8$=240 m. Some filtering and oversampling techniques [Kolb et al., 2012] can enhance the time resolution, but it is also possible to use the number of realizations as a strategy to improve the estimation.

The estimation error is reduced when the number of realizations is increased ($N \uparrow$). The estimation is carried out several times. Therefore, it is possible to improve the results even if bandwidth and SNR are not optimal.

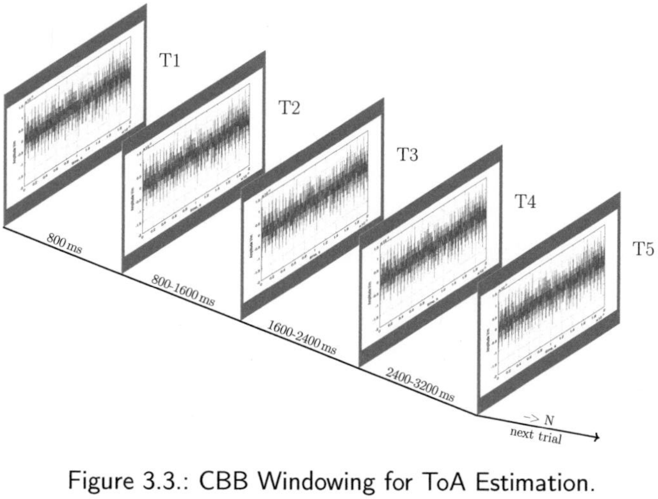


Figure 3.3.: CBB Windowing for ToA Estimation.

Suppose we define the $10^{th}$ strongest SC in the channel ($L = 10$) and set each realization using 800 samples, overlapping within the total number of 250,000. As indicated in Figure 3.3, the CBB signal is split into several slices; each one is a realization for the multipath estimation process. As shown in Figure 3.3 each estimation is done in a time-windows $T_i$, that defines each estimation frame.

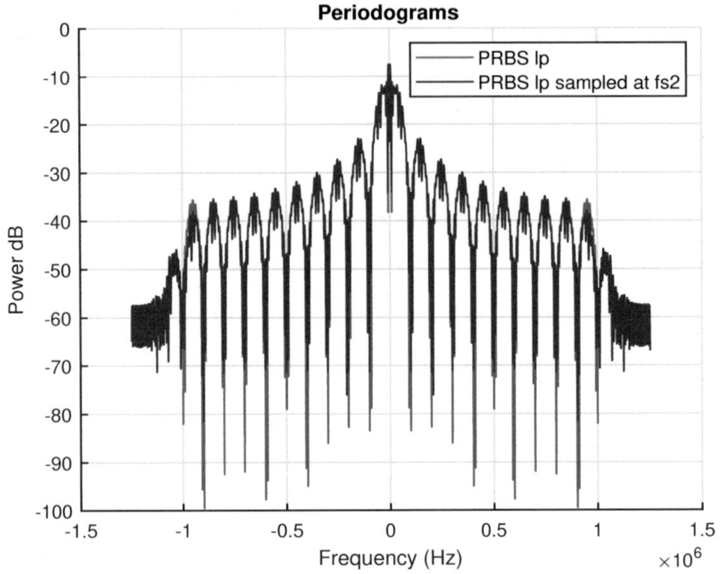


Figure 3.4.: CBB for CIR Estimation.

Let us analyze one example of the multipath estimation to compare the performance of blind CIR estimation with Multiple Signal Classification (MUSIC) and the one obtained with a classical convolution.

Assuming that a signal in baseband is a Pseudo-Random Binary Sequence (PRBS(13)), as defined in [ITU-T, 2000], with BW of 1 MHz in CBB, sampled using a $f_s = 1.25$ MHz, with a maximum duration of 20 $ms$, with an SNR of 10 dB. Figure 3.4 shows the periodogram of the signal received. The CIR is characterized by five SC as described by Figure 3.5, the delays are $\tau_i = [73, 250, 1114, 1202, 1364]ns$ and the amplitudes are normalized.

We also define the maximum number of the delay (or sampled frequency in the FFT) used in Equation (3.6) as $M = 28$, meaning that maximum delay considered in the multipath estimation is $\tau_{Max} = 28 \times 800ns = 2.210^{-5}$. The number of realizations can be defined by $N = 250000/100$=2500 realization.

Figure 3.5 shows the result of the simulation using MUSIC and deconvolution approach for extracting the multipath from the CBB signal shown in Figure 3.4. As described by [Botteron et al., 2004], MUSIC presents better resolution and the $4^{th}$ and $5^{th}$ peaks are difficult to extract using the given bandwidth. It is a typical case where a large number of realizations can enhance the estimation precision.

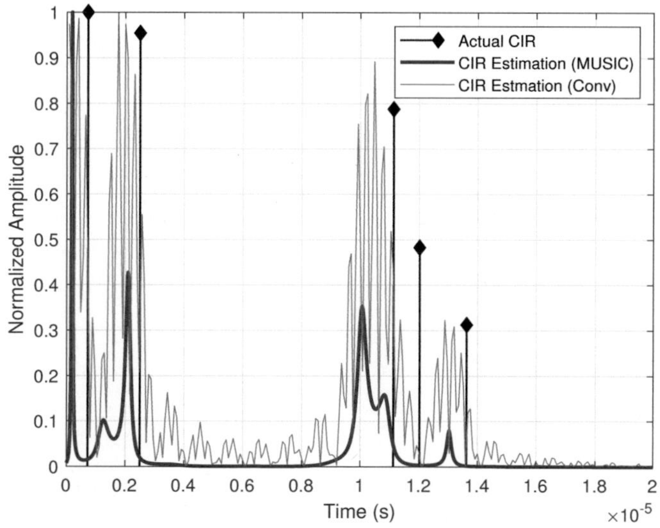


Figure 3.5.: Multipath Estimation using MUSIC and Convolution.

In short, this section studied the limits and the condition of multipath estimation using the signal received in each TDoA sensor. The lack of suitable SNR and BW affects estimation algorithms' ability to minimize estimation errors. The idea is to increase the number of realizations to improve the precision; in some cases, this is achievable, and in some others, it is more challenging. This strategy can reduce the error in some cases while using others, which might not be achievable.

## 3.4. Blind Estimation using SIMO FIR

The purpose of this section is to examine blind estimation approaches based on Channel Equalization. As described in the Figure 3.6, we can model the TDoA localization problem as a SIMO FIR system, [Liu et al., 1996], exploring the connection with the fingerprints localization approach.

The previous section shows that the channel impulse response is necessary for building our multipath-assisted fingerprints idea. The identification of unknown channel impulse response could be made with a training sequence. However, in our scope, it is not possible because the emitter is assumed non-collaborative.

A better way to solve this problem comes from the use of blind channel equalization. It is possible to estimate the channel response based solely on the channel output without using a training sequence using a blind channel equalization approach, [Sato, 1975].

A large number of data samples can improve the time resolution, even when the bandwidth and SNR are not optimal. The estimation can be set as an equalization problem using a class of subspace-based blind identification algorithms. Unfortunately, this method is limited by the nature of the problem at hand. The channels with defined SC over the DMC presents better results, [Moulines et al., 1995].

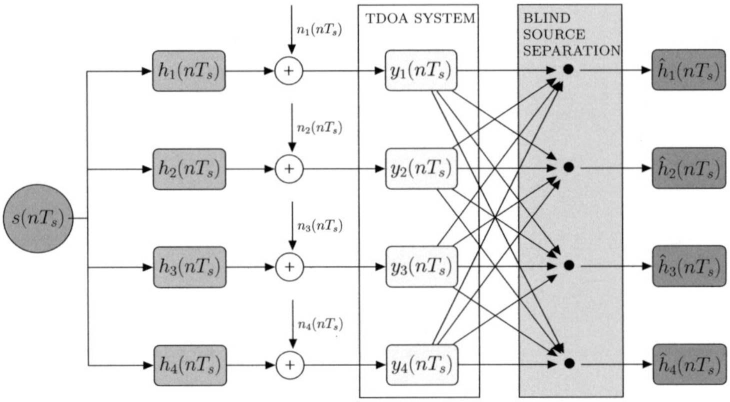


Figure 3.6.: Multipath Extraction with SIMO FIR.

Figure 3.6 shows the TDoA system as a discrete multichannel with FIR impulse responses $\{h_i(\cdot)\}$ driven by the same input $s(\cdot)$. The system output $y_i(\cdot)$ is related to $s(\cdot)$ and $h_i(\cdot)$ by:

$$y_i(k) = \sum_{j=0}^{L} h_i(j)s(k-j), \quad i = 1, \ldots, P. \tag{3.18}$$

$P = 4$ is the number of channels, and $L$ is the maximum order of the $P$ channels ($P \geq 2$ and $L \geq 1$). Let $N$ be the number of samples of the CBB signal taken in one slice. One observation or realization of the channel output is $y_i[k], i = 1, \ldots, P$, with the sample number, $k = L, \ldots, N$, meaning that the signal has some delay in arriving in each sensor.

The channels estimation of $\{h_i[k]\}_i^P$ is done using the noise-corrupted CBB received in each TDoA sensor without using higher order statistics.

The propagation channel from the transmitter to each receiver can be modeled as an FIR filter. Although the FIR equalization could also be applied to estimate the input $s(\cdot)$, this is out of our multipath approach.

We can use the formulation defined by [Chau, 2002], if the channel and the noise in each signal received has the following characteristics:

**(a)** The noise is complex normal and white with spectral density $\delta_n^2$.

**(b)** All emitter-sensor channels are linear time-invariant (LTI), with finite duration.

**(c)** There is at least one channel of length L (number of rays).

The channel with these features is classified as "identifiable," as described by [Moulines et al., 1995]. In this sense, the CIR can be identified uniquely if and only if the data matrix $\boldsymbol{Y}(L)$ is of rank $M(L+1)-1$. Second, the blind identification solution is unique if the sub-channel impulse response $h_i[k]$ and $h_j[k]$ do not share any common zero. These two conditions give sufficient conditions for blind identification of SIMO system.

Assuming the noise as a stationary Gaussian white random process, whereas the signals are deterministic unknowns, taking two CBB arriving in any pair of two noise-free outputs $y_i(k)$ and $y_j(k)$,

$$\begin{aligned} h_j(k) * y_i(k) &= h_j(k) * (h_i(k) * s(k)) \\ &= h_i(k) * \underbrace{(h_j(k) * s(k))}_{=y_i(k)} = h_i(k) * y_j(k), \end{aligned} \tag{3.19}$$

For SIMO systems with $P$ channels, there will be $P(P-1)/2$ output pairs. The term $*$, stands for convolution. Equation (3.19) can be described in matrix form as:

$$[\mathbf{Y}_i(L) - \mathbf{Y}_j(L)] \begin{bmatrix} \boldsymbol{h}_j \\ \boldsymbol{h}_i \end{bmatrix} = 0 \tag{3.20}$$

Remembering that if we have $L$ multipath components,

$$\boldsymbol{h}_p = [h_p[L], h_p[L-1], \ldots, h_p[0]]^T \tag{3.21}$$

, and

$$\mathbf{Y}_P(L) = \begin{bmatrix} y_P[L] & y_P[L+1] & \cdots & y_P[2L] \\ y_P[L+1] & y_P[L+2] & \cdots & y_P[2L+1] \\ \vdots & \vdots & \ddots & \vdots \\ y_P[N-L] & y_P[N-L+1] & \cdots & y_P[N] \end{bmatrix}. \tag{3.22}$$

Matrix $\mathbf{Y}_m(L)$ has a dimension of $(N-2L+1)$ by $(L+1)$. Using a compact notation to combine all the sensor channels in one single vector we have:

$$\boldsymbol{h} = [\boldsymbol{h}_p^T, \boldsymbol{h}_2^T, \ldots, \boldsymbol{h}_P^T]^T, \tag{3.23}$$

where $\boldsymbol{h}$ has a dimension of $M(L+1)$ by 1.

As shown in [Moulines et al., 1995] and [Chau, 2002], compact notation presenting all possible cross relations of the $i^{th}$ TDoA sensor is given as:

$$\boldsymbol{Y}^i(L) = \begin{bmatrix} \mathbf{0} & \cdots & \mathbf{0} & \mathbf{Y}_{i+1}(L) & -\mathbf{Y}_i(L) & \mathbf{0} & \mathbf{0} \\ \vdots & \vdots & \vdots & \vdots & \mathbf{0} & \ddots & \mathbf{0} \\ \mathbf{0} & \cdots & \mathbf{0} & \mathbf{Y}_m(L) & \mathbf{0} & \cdots & -\mathbf{Y}_i(L)) \end{bmatrix}. \tag{3.24}$$

For a TDoA system with $P = 4$ sensors, we would have

$$\boldsymbol{Y}^1(L) = \begin{bmatrix} \mathbf{Y}_2(L) & -\mathbf{Y}_1(L) & \mathbf{0} & \mathbf{0} \\ \mathbf{Y}_3(L) & \mathbf{0} & \mathbf{Y}_1(L) & \mathbf{0} \\ \mathbf{Y}_4(L) & \mathbf{0} & \mathbf{0} & \mathbf{Y}_1(L) \end{bmatrix}, \tag{3.25}$$

$$\boldsymbol{Y}^2(L) = \begin{bmatrix} \mathbf{0} & \mathbf{Y}_3(L) & -\mathbf{Y}_2(L) & \mathbf{0} \\ \mathbf{0} & \mathbf{Y}_4(L) & \mathbf{0} & -\mathbf{Y}_2(L) \end{bmatrix}, \tag{3.26}$$

$$\boldsymbol{Y}(L) = \left.\begin{bmatrix} \begin{matrix} \boldsymbol{Y}^1(L) \\ \vdots \\ \boldsymbol{Y}^{P-1}(L) \end{matrix} \\ \underbrace{\hphantom{\boldsymbol{Y}^{P-1}(L)}}_{P \quad Blocks} \end{bmatrix}\right\} \frac{P(P-1)}{2} \quad Blocks. \tag{3.27}$$

In this structure, $Y(L)$ has dimension $(N-2L+1)\frac{P(P-1)}{2}(L+1)$. The $\frac{P(P-1)}{2}$ matching equation to estimate the CIR from the CBB is obtained by solving:

$$\boldsymbol{Y}(L)\boldsymbol{h} = \boldsymbol{0} \tag{3.28}$$

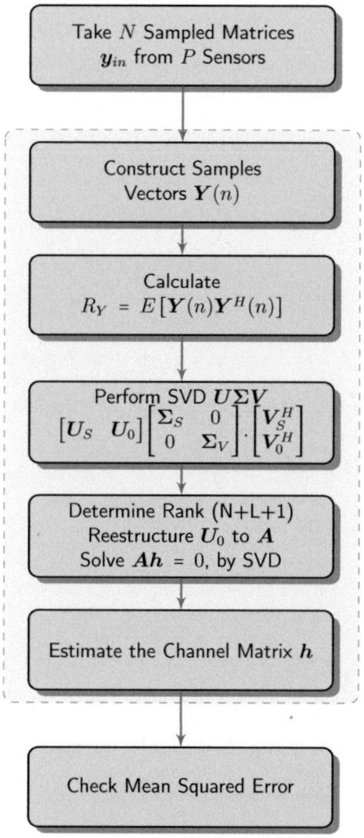


Figure 3.7.: SIMO FIR Channel Estimation.

Figure 3.8 shows the SVD approach to extract the multipath information as a SIMO FIR channel estimation problem. In Equation (3.28), vector $\boldsymbol{h}$ is the null space for the case of noiseless. The expression for the LS approach is:

$$\|\boldsymbol{Y}(L)\boldsymbol{h}\|^2 = \boldsymbol{h}^H\boldsymbol{Y}^H(L)\boldsymbol{Y}(L)\boldsymbol{h} = 0 \tag{3.29}$$

$\boldsymbol{A}$ is also called matrix of measurements, and it has the following structure:

$$\boldsymbol{A} = \boldsymbol{Y}^H(L)\boldsymbol{Y}(L). \tag{3.30}$$

$\boldsymbol{A}$ matrix can be solved by:

$$\boldsymbol{h}^H\boldsymbol{A}\boldsymbol{h} = 0. \tag{3.31}$$

Equation (3.31) can be solved via SVD. If noise is added to the model, the equation becomes non-homogeneous, giving wrong estimations. In this case, the LS approach can once more be used.

$$\min_{\hat{h}} \left\|\boldsymbol{h}^H\boldsymbol{A}\boldsymbol{h}\right\|^2 \tag{3.32}$$

Figure 3.8.: Multipath Extraction with SIMO FIR

So the estimated CIR is the eigenvector corresponding to the smallest eigenvalue of the measurements matrix $\boldsymbol{A}$.

Figure 3.9 shows the SIMO FIR estimation simulation where we have considered only four multipaths in each channel ($L = 4$). The estimation procedure can also combine the algorithms estimated in Section 2.3.

In blind estimation, it is necessary to define the length of the channels in each TDoA Sensor. The knowledge of the model order is critical to blind identification algorithms as it is to most parametric estimation methods.

It is possible to assume that the lengths of the channels are known as a priority. In this case, it is relevant to recall the study results of [Landmann, 2008], showing that in outdoor scenarios, the identifiable multipath components are usually five or seven.

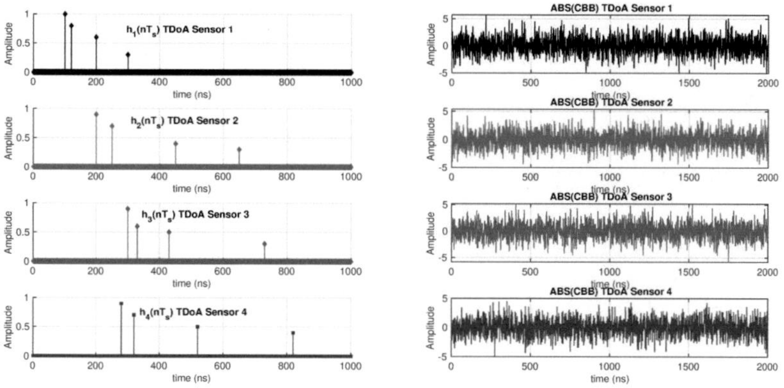


Figure 3.9.: SIMO FIR Multipath estimation using CBB in TDoA Sensors.

Nevertheless, there are objective criteria to define the model order in the estimation, either the minimum description length (MDL) [Huang and Benesty, 2007] or the Akaike information criterion (AIC), [Akaike, 1974].

In summary, it is possible to model the TDoA system as a SIMO FIR equalization problem to estimate the multipath. There are several constraints related to the SNR and the sensor BW, between the actual rays and the ones produced by the signal processing. The idea of applying channel subspace (CS) will also require more significant computational effort in the sensor node. In case of the estimation is performed in the fusion node, it is necessary to transmit long recordings of sampled CBB signals that, in most cases, are not feasible in an outdoor scenario.

## 3.5. Multipath Extraction using Volume Cross-correlation

This section shows the VCC proposed by [Shi et al., 2016] as an innovative signal processing strategy to extract multipath information from TDoA systems. In Section 2.4.3, we stated that in localization systems based on TDOA [Zekavat and Buehrer, 2011], [Meng et al., 2016], [He et al., 2017], [Kolb et al., 2012], the time differences are the lags in the peak of the cross-correlation function between two complex base-band

(CBB) signals received. This information can also be extracted from the peak in the Volume Cross-Correlation function.

[Shi et al., 2016] showed the derivation and the modeling of the VCC for fast and slow-changing channels. We have adopted only the formulation for slowly changing channels once we considered slow velocity objects and almost a static scenario.

The conventional wireless communication signals (DS-CDMA, OFDM, QAM, etc.) have channel delays caused by different distances between receivers or reflective objects that do not have high velocities; therefore, channel delays can be assumed constants.

$$\boldsymbol{y}_i^{(j)} = \boldsymbol{G}_i \boldsymbol{\alpha}_i^{(j)} + \boldsymbol{w}_i^{(j)}, \qquad j = 1, 2, \ldots, \tag{3.33}$$

where $j$ indicates different observations at different times or snapshots. Although the channel coefficient vector $\boldsymbol{\alpha}^{(j)}$(j) is fluctuating with different $j$, the subspace structure is determined by a matrix $\boldsymbol{G}_i$, will remain almost unchanged. We call this category of signals *the slowly changing subspace signal*, meaning that the subspace structure in Equation (3.33) changes slowly with time and can be treated as invariant during the observation interval.

There are several variations of the cross-correlation concept to obtain time-differences, [Brutti et al., 2008]; nevertheless, the ability to resolve close paths in a wireless channel with a dense multipath effect is still an open issue. We reproduce the theoretical analysis and simulation to show how this approach can be integrated into our multipath exploitation scheme.

Figure 3.10 shows the difference between VCC and the SIMO FIR. Now the CBB signal from each sensor is sent to the central processing node to perform the TDoA. It is possible to extract the multipath information from the VCC directly. We intend to take the principal peak in the correlation function and other maximums related to the multipath composition in a given sensor pair with this method.

We have concentrated the studies in the VCC because of the superior performance to identify each multipath TDOA [Shi et al., 2016]. The initial idea of the VCC is to decompose each CBB signal and perform the cross-correlation with the signal basis function. This multi-dimensional cross-correlation approach can resolve multipath TDOAs.

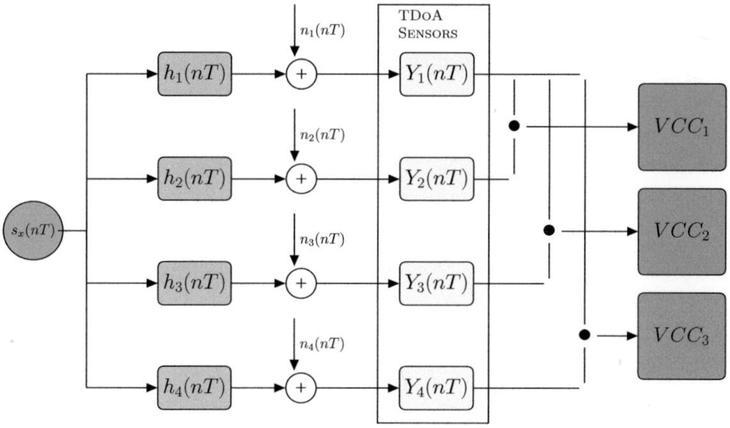


Figure 3.10.: Multipath extraction with VCC.

The name "Volume" stands for the geometric high dimensional parallelotope [Valaee and Kabal, 1995] that is produced by spanning the column vectors of the signal matrices. After this operation, the subspace is extracted from noise through singular value decomposition of the data matrix.

The VCC is related to the Principal Component Analysis (PCA), where the inputs are estimated using a subspace basis. The TDOA estimation is the spam position of the $\Delta_i$ that are zeros of the VCC function (or equivalently, peaks of its reciprocal).

The signals received in Sensors S1 and S2 can be modeled as:

$$y_1(t) = \sum_{l_1=1}^{L_1} \alpha_{1,l_1} s_t(t - \tau_{1,l_1}) + n_1(t), \tag{3.34}$$

$$y_2(t) = \sum_{l_2=1}^{L_2} \alpha_{2,l_2} s_t(t - \tau_{2,l_2}) + n_2(t), \tag{3.35}$$

In Equation (3.34) and Equation (3.35), $\alpha_{1,l_1}$ and $\alpha_{2,l_2}$ are the channel coefficients associated to the $l_1^{th}$ (or $l_2$) rays, related to the sensor-pair $2-1$; the path delays are $\tau_{1,l_1}$ and $\tau_{2,l_2}$.

The number of paths detected in each channel are defined by $L_1$ and $L_2$, $n_1(t)$ and $n_2(t)$ are the noise in each sensor, that can also be estimated by [Huang and Wu, 2007] or [Akaike, 1974].

As explained in [Shi et al., 2016], the VCC gives not only the time delay estimation between two signals; it is also possible to extract the combination of the multipath, that are characterized by other peaks. The multiple TDoA) caused by several combinations between the delays in the sensor are represented by:

$$\Delta\tau_{l_1,l_2} := \tau_{2,l_2} - \tau_{1,l_1}, l_1 = 1, \dots, L_1, l_2 = 1, \dots, L_2, \tag{3.36}$$

Assuming the signals received in Sensors 1 and 2, in matrix form, are represented by $\mathbf{S}_1 \in Cn \times d_1$ and $\mathbf{S}_2 \in Cn \times d_2$. The VCC using these two matrices is:

$$\text{VCC}(\boldsymbol{Y}_1, \boldsymbol{Y}_2) := \frac{\text{vol}_{d_1+d_2}([\boldsymbol{Y}_1, \boldsymbol{Y}_2])}{\text{vol}_{d_1}(\boldsymbol{Y}_1)\text{vol}_{d_2}(\boldsymbol{Y}_2)}, \tag{3.37}$$

$[\boldsymbol{Y}_1, \boldsymbol{Y}_2]$ is the concatenating matrix, produced joining the columns of matrices $\mathbf{S}_1$ and $\mathbf{S}_2$ into one. The term $\text{vol}_d(\mathbf{S})$ represents the geometrical volume of matrix $\mathbf{S} \in C^{n \times d_2}$ with dimension $d(d < n)$, [Ben-Israel, 1992] and [Shi et al., 2016].

Algorithm 1 shows the pseudo-code for the VCC calculation. An in-depth derivation of the subspace angles is available in [Miao and Ben-Israel, 1992], and the volume produced by a linear matrix can be found in [Ben-Israel, 1992]

In Algorithm 1, the dimension of the signal subspace ($L_i$) gives the number of rays in the CIR. The number of multipath in the VCC or the model order can be estimated using the approach of [da Costa et al., 2011], based on the observation in the real measurements in the scenario considered. The VCC function of two signals can give the time-delay estimation (TDE) and the combination of each of their multipath components.

Figure 3.11 shows the result of the TDOA estimation using VCC and MUSIC in a synthetic Channel Impulse Response. An FM modulated signal with BW of 1MHz, sampled at 1.25 MHz and upsampled to 12.5 MHz the time resolution of $\Delta_t = 80ns$. As indicated, the five maximum in the delays were at $[75, 112, 800, 925, 1000]\mu s$, the MUSIC could not identify the $2^{nd}$ and the $4^{th}$ peaks, and the VCC could identify them all.

In Figure 3.12, we observe the result of performing VCC several times in a long recording of the signal that arrives in two distinct TDoA sensors. As we can observe, the realizations using an integration time of 10 ms show that the maximum and the other peaks in the VCC do not change during the time. It indicates that the multipath patterns

**Algorithm 1:** Peaks in VCC TDoA processing, based on [Shi et al., 2016].

**Input** : CBB signal in Sensor 1 and 2 ($\boldsymbol{Y}_1$ and $\boldsymbol{Y}_2$)
**Output:** Peaks in VCC ($\Delta\tau$ and $\alpha$)
**for** $\Delta\tau \in [(-N+1)T_s, (N-1)T_s]$ **do**
  Obtain multiple delayed data (for $j = 1 \ldots m$ );
  $\mathbf{Y}_1^{(j),[\Delta\tau]} = [Y_1^{(j)}(0 \cdot T_s - \Delta\tau) \ldots Y_1^{(j)}((N-1)T_s - \Delta\tau)]^T$;
  Obtain observation not delayed data;
  $\mathbf{Y}_2^{(j),[\Delta\tau]} = [Y_2^{(j)}(0 \cdot T_s) \ldots Y_2^{(j)}((N-1)T_s)]^T$;
  Obtain Covariance Matrices $\hat{\boldsymbol{R}}_1^{[\Delta\tau]}$ and $\hat{\boldsymbol{R}}_2$;
  Perform eigenvalue decomposition of $\hat{\boldsymbol{R}}_1^{[\Delta\tau]}$ ( also for $\hat{\boldsymbol{R}}_2$ );
  $\hat{\boldsymbol{R}}_i = \boldsymbol{U}_{i,s}\boldsymbol{\Lambda}_{i,s}\boldsymbol{U}_{i,s}^H + \boldsymbol{U}_{i,n}\boldsymbol{\Lambda}_{i,n}\boldsymbol{U}_{i,n}^H$ ;
  Estimate the dimensions of signal sub-spaces $L_1$ (and $L_2$) based knowledge of the eigenvalues;
  Extract the signal sub-spaces $\boldsymbol{U}_{1,2,s}^{[\Delta\tau]} \in C^{N \times L_1}$ from the eigenvectors corresponding to the $L_{1,2}$ largest eigenvalues of $\hat{\boldsymbol{R}}_1^{[\Delta\tau]}$ (and $\hat{\boldsymbol{R}}_2$);
  Calculate the reciprocal of VCC function: $\rightarrow \mathbf{r}_{vol}(\Delta\tau) := \frac{1}{\mathbf{Vol}_{L_1+L_2}([\boldsymbol{U}_{1,s}^{[\Delta\tau]}, \boldsymbol{U}_{2,s}])}$;
  Find the Peaks in: $\mathbf{r}_{vol}(\Delta\tau)$
**end**

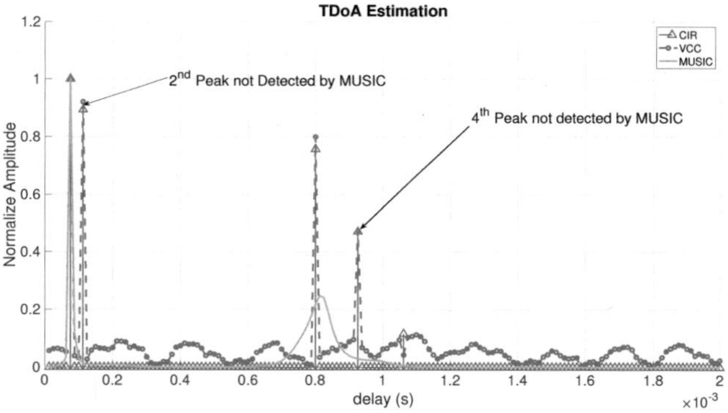


Figure 3.11.: Performance of VCC, MUSIC and GCC in multipath extraction from TDoA.

caused by the obstacles are embedded in the VCC function.

In a nutshell, the Volume Cross-Correlation can be used as a super-resolution TDOA estimation technique, which presents better performance than MUSIC. The usual TDoA system, as described in Section 2.3, takes only the peak in the cross-correlation function to estimate the time delay. The novelty explored here was how to use the other maximums as multipath fingerprints for localization. The VCC, based on subspace and principal angles

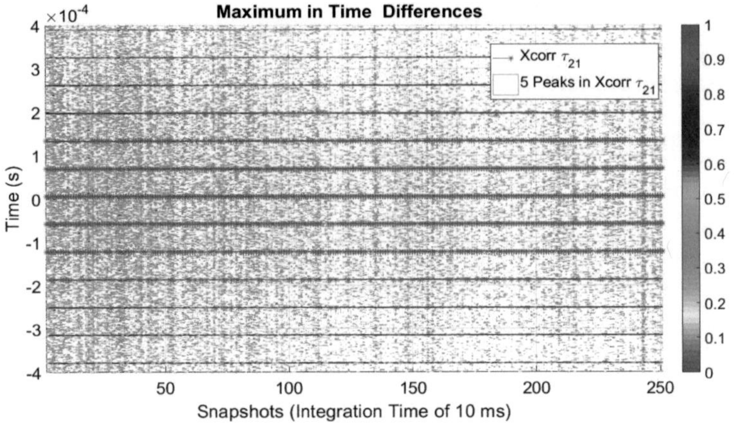


Figure 3.12.: Multipath extraction in several VCC realizations.

between the signal considered, showed better results than the MUSIC algorithm.

## 3.6. Concluding Remarks

Channel Estimation is a central topic in several areas of wireless communication. This chapter has explored the critical aspects of the CIR in the localization framework.

We have investigated blind estimation techniques based on super-resolution and SVD and the multipath patterns buried in the VCC function. These methods are similar in complexity, sensitivity to noise, and performance to estimate closely spaced multipath components.

As described in this chapter, the multipath fingerprints can be extracted using the CBB directly or using the VCC. In the case of CBB, the computing requirements for locally processing the signal to obtain the ray's information is a drawback in the implementation strategy. Therefore, the VCC approach seems to be a better approach for implementation because the cross-correlation step is already required in the position processing.

Finally, the time resolution to reduce the errors in the multipath feature extraction depends on both the SNR and the BW. We have shown that it might be possible to perform several realizations. The multipath resolution can be increased by repeating the process repeatedly, applying statistical methods for CIR estimation.

We have explored with simulations the ability of several realizations in reducing estimation errors. This averaging approach is feasible if the scenario is static or with slow changes during the measurement period.

The fundamentals of TDoA system, investigated in Chapter 2, including the practical aspects of the restriction in BW, together with the multipath pattern extraction from the signals received in each sensor explained here, are the foundations for the multipath fingerprints localization. With these two building blocks, it is possible to assemble the positioning fingerprints framework using a multipath estimation approach.

# 4. Multipath Exploitation with Ray Tracing

This chapter aims to explain the procedures adopted to generate the radio frequency multipath fingerprints used in our method. The Ray Tracing (RT) simulation was applied to the desired outdoor environment to estimate the CIR as a characterization of the wireless propagation channel, a fundamental issue in localization systems.

Therefore, we provide an overview of the essential aspects of the RT multipath exploitation idea for emitter positioning. RT has been widely used to predict the CIR both in outdoor and indoor scenarios. Classic RT tools take into account Line Of Sight (LOS), specular reflection and diffraction. Typically, RT can simulate the propagation, relying on different types of models: empirical, statistical, and deterministic.

In our approach, the properties of the electromagnetic rays follow a Geometrical Optics (GO) procedure for the interaction effects with the environment. The central idea is to enhance TDoA systems using computer simulations to extract positioning characteristics of the multipath rays in a given region.

The method requires knowledge of the locations of the TDoA sensors, a scenario plan-view map with buildings description, and characteristics of the outdoor environment. The approach can accommodate both cooperative and non-cooperative sources. However, it is necessary to include some parameterized information about the frequency range and link parameters, such as power output and polarization, to feed the simulation engine.

The chapter has five sections plus the partial concluding remarks. In the first, we provide a survey of the related research, emphasizing the studies carried out in the multipath extraction using RT, multipath exploitation, and the key aspects used in our localization framework. In the second part, we present the theoretical foundation of RT simulation tools, where we review the general aspects instead of being attached to some specific commercial RT implementation.

In the third section, we indicate the main drawbacks when using RT. In our approach, the outdoor CIR estimation is usually imperfect; some of the predicted rays can not be measured with the restrictions explained in Chapter 3. Bandwidth and polarization limitations of the RT are also discussed.

Following the investigation, this chapter shows a procedure to build a CIR and also a TDoA database using RT results. Consequently, we explain how we can produce synthetic data to compare the measurement of the multipath extracted from the TDoA systems explored in Chapter 3.

Finally, we propose a framework to generate not only the CIR fingerprints but also the VCC synthetic characterization of the scenario. Thus, we perform a general description scheme for a cooperative relay network. Following, we show some examples of the RT multipath fingerprints produced from a typical suburban scenario.

## 4.1. Motivation and Related Work

This section highlights the significant studies to understand the proposed multipath fingerprints. We briefly explain the related works that also tried to use RT to characterize outdoor scenarios.

In previous chapters, we have outlined TDoA practical aspects in Chapter 2 and explained multipath concepts used in localization systems in Chapter 3. Now, we focus on the multipath exploitation idea.

The primary motivation of going deeper in the RT technique is to elucidate in which circumstances it is possible to produce multipath fingerprints as a mixture between real measurements and simulated ones.

[Baker and Pratt, 2013] showed a geolocation approach applicable to multipath environments, where they fused RT simulation with the 2D map, including information of the buildings to produce multipath patterns for ambiguity reduction in geolocation.

[Mani, 2012] proposed a parametric procedure to include penetration, diffuse scattering, and contributions by vegetation into empirical, statistical, and deterministic models for CIR. The author has presented simulation results and a comparison with real measurements to validate the RT models.

There is a study in [De Groot et al., 2014] of localization and tracking using a hybrid fingerprint database that includes RSS, TDoA and AoA. The database is created with OpenStreetMap building data to model the area of interest and estimate the origin of incoming signals. He showed simulations using three receivers in a 10m square grid, achieving a localization accuracy of 15m.

[Hussain, 2017] investigates an RT based on dominant path computation considering the geometry of the environment, proposing a new method to accelerate the ray computations using the intra-visibility matrix approach. [Naseri and Koivunen, 2016] developed a two-dimensional joint localization and mapping method that uses convex optimization and multipath delay to provide a reliable indoor positioning.

[Fuschini et al., 2015] performs an overview of traditional RT techniques and analyses future applications, exploring the diffuse scattering and multidimensional characterization. Furthermore, they offer a real-time ray-tracing scheme to multiple-input multiple-output (MIMO) capacity assessments based on tests carried out at the University of Bologna.

[Phelan et al., 2012] proposed the idea of Multipath Characteristic Database (MCD) containing the multipath signatures for each possible transmitter location in an area of interest using RT. They developed a multipath signatures scheme by minimizing the distance between estimated and simulated multipath feature vectors. [Kikuchi et al., 2006] combined the spatial characteristics estimated from data measurement and ray-tracing (RT) for determining the position of an unknown mobile terminal. A blind estimator was proposed in this reference, using a single base station with an array of sensors in a multipath environment.

[Porretta et al., 2003] explained the idea of using a localization system based on RT assisted simulation. This original algorithm used a single base station to locate mobile terminals in cellular networks, adding information about the environment in the base station neighborhood. [Martinsson and Johnsson, 2002] investigated the deterministic indoor channel model characteristics using a RT simulator to include the physical properties of construction materials.

[Gowdayyanadoddi et al., 2015] presented an application of RT to predicts Doppler differences in the GPS receiver in NLOS, showing that the accuracy of the method using measured Doppler differences from real data collected in a city environment.

[Gentner et al., 2016] offers comprehensive examinations of Multipath components (MPCs) for localization and tracking in a typical urban scenario. It was analyzed the lifetime of the multipath rays and the propagation process for locations based on scattering objects. Finally, [Vin, 2014] proposed a localization method for urban environments based on multipath fingerprints, where a deterministic RT channel model is used to locate a mobile emitter in a limited area.

[Lawton and McGeehan, 1994] investigate the ability of ray launching techniques to evaluate reflected, transmitted, and diffracted rays from a simplified description of an environment. The authors used simulations and real measurements to demonstrate the model's ability to predict typical RMS delay spread values.

[Roivainen et al., 2016] presented a parametric three-dimensional (3-D) geometry-based stochastic radio channel model (GSCM) based on measurement campaigns. He compared the model with real measurements showing a good match between estimated and modeled multipath components (MPCs).

[Wielandt and Strycker, 2017] examined a multipath propagation fingerprints exploitation scheme to enhance the accuracy of AoA localization systems. An RT was used for indoor position determination, and the results were evaluated experimentally, presenting a fingerprinting approach for matching the AoA measurement vector to a set of reference vectors.

We have collected here some researches that have focused on the use of the ray-tracing simulation tool to obtain the CIR; other investigations used RT to produce a database of possible multipath. Our contribution to this field was expanding the concepts presented herein to produce a synthetic database of multipath patterns, either using the CIR function or using the cross-correlation function itself.

In summary, the related research let us conclude partially that RT can be applied to produce a multipath pattern to be used on outdoor scenarios. Furthermore, several studies have shown strong agreement between measurements and simulations. Therefore, we have taken the central ideas presented here and used them as a foundation to build a multipath fingerprint for enhancing TDoA systems.

## 4.2. Ray Tracing Techniques

This section explains the fundamental aspects of RT as a deterministic 3D modeling of wave propagation, describing how to integrate it in our multipath fingerprints approach.

In urban areas with several obstacles, the standard empirical propagation models do not give suitable coverage predictions. Therefore, it is necessary to use deterministic models based on ray optical in combination with diffraction theory [Yun and Iskander, 2015].

Maxwell's equations can give the full wave equation describing the multipath propagation. However, they present a high degree of complexity in an outdoor scenario.

The RT algorithm can simplify the problem by using a high-frequency approximation, where the wave-fronts are plane waves, following the principles of GO, Uniform Theory of Diffraction (UTD) and Physical Optics (PO) [Balanis, 2012].

It is a suitable method to characterize the CIR in large and complex environments like our outdoor TDoA localization problem. A single ray can interact by reflection, diffraction, or scattering, depending on the obstacle size and position.

The RT computational cost is lower than the full-wave solvers like Method of Moments (MoM); the memory requirement depends on the desired discretization, and it does not increase with frequency. The accuracy of the model also depends on the availability of exact geometry in the database.

It is important to differentiate RT from Ray Launching (RL) propagation models. Both methods can compute the CIR using the Geometric Optics approximation [Deschamps, 1972], the difference lies in the algorithm used to determine the paths that connect the emitter to the receivers [Gschwendtner et al., 1995].

In RL, the algorithm launches rays in all possible angles (azimuth and elevation) that are discretized into small angular increments. The field strength is summed up at all potential receiver regions when energy falls below a minimum, or a given maximum number of interactions is achieved.

The problem of this method is the necessity to establish a region surround the receiver to collect suitable rays to perform the CIR. The definition of this region turns out to be a disadvantage of the scheme. If the area is too small, some relevant rays may not be collected; if it is too large, some unnecessary ones may be considered.

Even with these limitations, the RL can be used for point-area coverage problems, but it is not very suitable for point-to-point prediction.

Figure 4.1 shows a simulation where we have pointed out the collecting region surrounds a receiver station. The size of this region defines the number of rays considered. Thus, we can observe how diffraction, reflection, and scattering effects are taking into account.

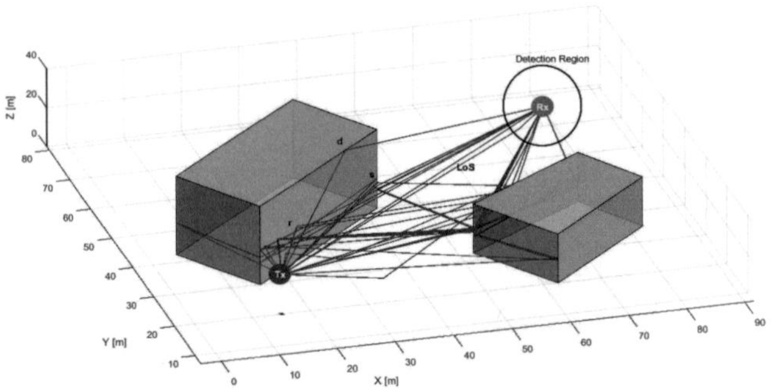


Figure 4.1.: Collection Region.

RT defines the path rays as a point-to-point algorithm computation. The complexity depends on the number of buildings and the prediction area.

The algorithm looks for valid ray paths between the transmitter and the receiver station, making a point-to-point connection. The electromagnetic material properties used in the wave attenuation defines if a ray is valid or not.

In the case of coverage evaluation, the RT needs to define several receiver points in all the prediction areas. In this case, the computation time increases with the dimension of the prediction area. Therefore, for each transmitter-receiver (Tx-Rx) combination, there is a CIR that defines the time delays for the specular components.

In RT, the ray trajectory needs to be set entirely, including the number and position of each interaction. Therefore, it is mandatory to apply a visibility algorithm to identify valuable surfaces on the scenario [Gschwendtner et al., 1995].

The core of ray-based algorithms is the definition of each path in the virtual scenario, which can be implemented using the Shooting and Bouncing Ray (SBR), the image theory, or a hybrid approach. All three techniques will be explained in the following subsections.

### 4.2.1. Shooting and Bouncing Ray Technique

The RL usually implements a SBR technique, where the rays start at the emitter position and follow arbitrary directions until a stopping condition is met. Thus, there is no pre-processing step in the computation, and the complexity grows linearly with the number of iterations.

In the SBR technique, the rays are launched from the transmitter position in all directions. Only the waves that reach the receiver are collected.

If an incident ray hits a wall and reflection occurs, the computation is done by using the GO or the UTD techniques, where rays are straight lines in all possible directions covering the whole space.

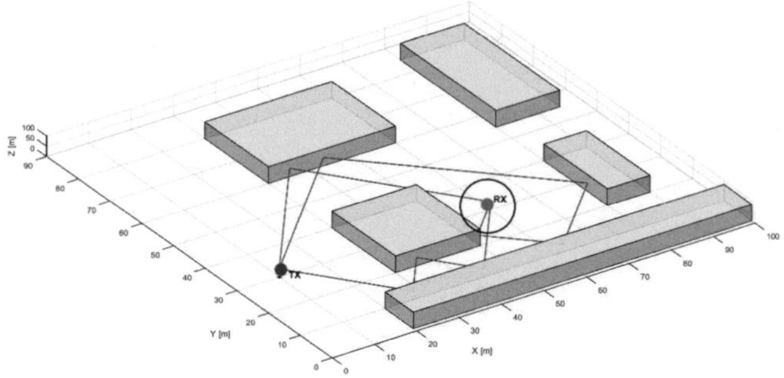


Figure 4.2.: Shooting and Bouncing Ray Technique.

Figure 4.2 shows how the algorithm works. The rectangular blocks are considered buildings. The circle surrounding the receiver is the "collecting" region, where the rays are considered valid in the simulation.

These rays form a uniform pattern and spread across the entire surface. An interaction between the beam and the object is characterized either by reflection or diffraction.

### 4.2.2. Method of Images

The method of images considers obstacle walls as mirrors, where all planes are taken into account for generating the images of a source. In addition, the images are virtual emitters acting as a secondary source for reflection points. Figure 4.3 shows the image method and the virtual emitter approach, with the help of which multiple ray reflection paths can also be determined in a recursive procedure.

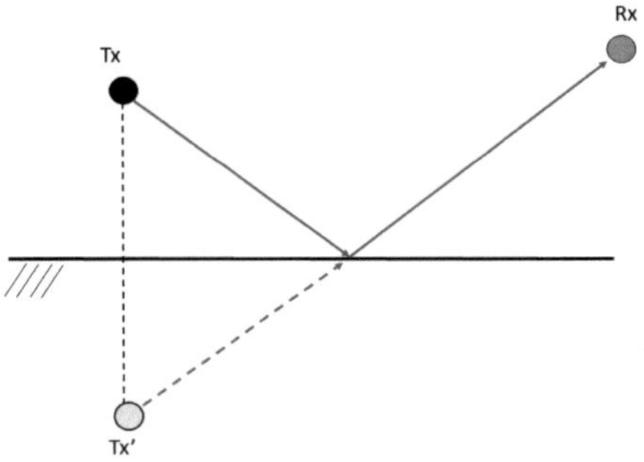


Figure 4.3.: Method of Image in RT.

The image theory can not be applied below the ground plane because it represents a conducting plane where the antennas are mounted near it. Another way of calculating the reflected ray path is to calculate the intersection point Q by the transmitter's data (Tx) and the receiver (Rx).

### 4.2.3. Hybrid Method

The described two methods have some disadvantages. Among these, some of the reflection points might be located outside the visible region. Another disadvantage is that the SBR technique has lower accuracy for tracing the path of each multipath component.

However, by combining both methods, it is possible to achieve better results. The hybrid method tries to find visible surfaces and use them to produce valid rays joining SBR and image theory to calculate the interaction points. Modern RT tools usually implement this hybrid approach by default to achieve an accurate ray path. Figure 4.4 shows the result of this hybrid approach.

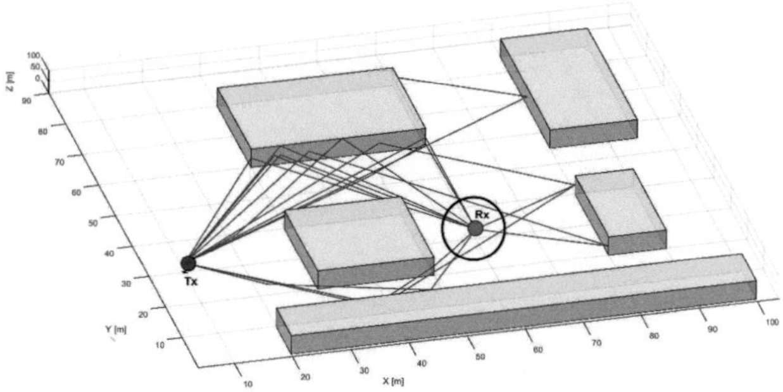


Figure 4.4.: Hybrid approach: SBR and Image Theory.

The ray-tracing technique usually employs the "tree of rays" scheme as a pre-processing step to traverse all faces and edges of an object. Then, the method recursively sets up to find the primary or secondary rays in all simulation domains.

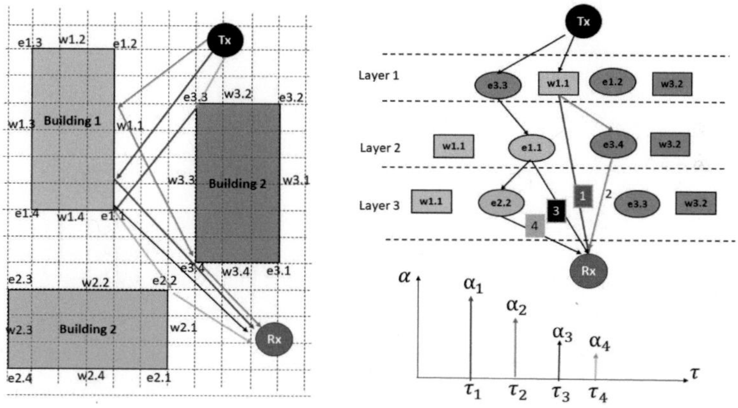


Figure 4.5.: Visibility Tree, adapted from [Fuschini et al., 2008].

There are several approaches to obtain the correct ray trajectory in the discretized simulation scenario. An in-depth study carried out by [Yun and Iskander, 2015] and [Larew et al., 2013] explores the algorithms to form the visibility tree representing the visibility relationships among the interacting objects and the transmitter and the receiver. These studies also explain some dimension reduction techniques to speed up simulations.

### 4.2.4. Visibility Tree in RT

RT explores the idea that visibility between the position in the simulation domain and the walls and edges remain constant in the environment. A pre-processing step computes all possible interactions in advance to speed up simulations; consequently, a visibility tree is produced.

This pre-computation is a central part of RT, which calculates all the ray segments present between walls and edges. It also computes the multiple interactions happening in the simulation domain. For example, visibility refers to the existence of a radio propagation path between two objects. Figure 4.5 shows the internal implementation of the RT interactions, in which the path first represents the objects in the Tx-Rx connection. Then, the tree structure is created to perform the calculations.

Figure 4.5 also shows the schematic representation of walls and edges in the ray-tracing simulation. With the output file of ray tracing, it is possible to search for the given positions where the reflectors bounce off the rays before arriving at the sensors. It works similarly to extracting the "field of vision" in an optics ray. The view tree contains all relevant visibility information for the actual geometrical tracing of rays in a given environment.

Visibility relations alone are not sufficient to do the RT calculation; hence other necessary steps, as specified in [Fuschini et al., 2008], should be used to accelerate the final computation. Furthermore, we may include other obstacles like vegetation, pedestrians, and vehicles if simplified in edges and faces.

Remember that the interaction number is limited in the pre-processing. There is no guarantee that the rays starting from the receiver's terminal are the same rays transmitted from the transmitter side. Visibility relations do not correspond to exact geometrical information. Therefore, after the view tree is constructed, other steps are necessary for the actual computation of both the ray geometry and field.

It is also relevant to point out that the RT is not necessarily reciprocal and depends on the defined collecting region and the SBR strategy, where the paths that connect the emitter-receiver may sometimes differ from the receiver emitter. The bidirectionality is a new class of RT tool investigated by [Taygur et al., 2018] to cope with the antenna coupling problem. The idea is to overcome the limitations of the reception spheres in the traditional methods, where both transmitter and receiver are used for ray launching.

The transfer function between two antennas is obtained by evaluating the reciprocity integral on this interaction surface by a particular numerical integration procedure.

## 4.3. Modeling of Ray Interaction in RT

This subsection explains the main aspects of the ray-object interaction implemented in the RT approach. It is relevant to analyze some fundamental elements of a generic and standard RT, to understand the proposed multipath fingerprints approach.

Buildings and other outdoor obstacles in the environment can be either extracted using a Light Detection And Ranging (LIDAR) approach [Rottensteiner and Briese, 2002] or taking it from some open-source database.

The OpenStreetMap project [Haklay and Weber, 2008] provides a collective database with scenario data, including buildings and vegetation fields. Sometimes, it is necessary to enhance the free database with updated information or even simplify some obstacles. For example, Figure 4.6 shows the buildings in Ilmenau, Germany, including the lakes, vegetation, and elevation profile.

Following the Geometrical Theory of Diffraction (GTD) from [Balanis, 2012] and [Fuschini et al., 2008], outdoor propagation can be modeled with some approximations. To model the complete full-wave solution is demanding. The propagation mechanism in outdoor environments can not be easily explained because different types of obstacles affect the link. Typically the receivers are surrounded by objects, including buildings, vegetation like trees and plants, pedestrians, and vehicles.

The actual electromagnetic characteristics of the object in the scene are not entirely described and sometimes are not available. Therefore, practical simulations using RT are performed using an approximation for the shape and the material obstacles. The RT can use the database either in two-dimensional or three-dimensional regions.

To effectively understand the RT tools, it is essential to investigate the fundamentals of electromagnetism of Geometric optics and the General Theory of Diffraction. A more complete and more in-depth derivation can be found in [Fuschini et al., 2008]. The

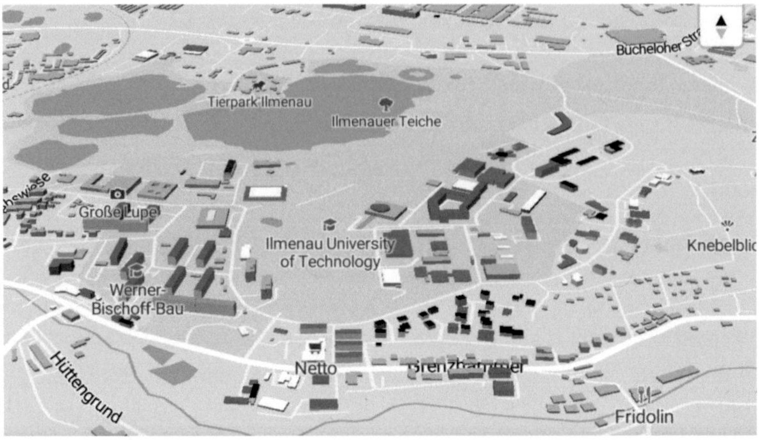


Figure 4.6.: Open-source Scenario Information [Haklay and Weber, 2008].

analysis presented here clarifies how the tool calculates the different iterations that the rays suffer in a multipath environment.

Thus, we can represent the free-space LOS generated in a point $i$ in spherical coordinates by:

$$\boldsymbol{E}_T(r,\theta_T,\phi_T) = I_T\sqrt{\frac{Z_T.\eta.g_T(\theta_T,\phi_T)}{16\pi}}\frac{e^{-j\beta_r}}{r}\boldsymbol{P}(\theta_T,\phi_T) = \boldsymbol{E}_{T0}\frac{e^{-j\beta.r}}{r}, \tag{4.1}$$

where $\boldsymbol{E}_T$ is the total electric field in spherical coordinate system, $Z_t$ is the antenna impedance, $I_T$ is the current induced on the antenna feeding terminals, $g_t$ is the dimensionless antenna gain, $\eta$ is the intrinsic impedance of the medium, $\beta$ is the wave-number and $\boldsymbol{P}(\theta_T,\phi_T)$ is the antenna polarization vector.

As shown in Figure 4.7, several SC are selected as valid rays by the RT algorithm. The model can include reflection, diffraction, scattering, or a combination of all of them. Thus, GTD approximation allows us to model the electromagnetic effects of each ray to estimate all the CIR parameters of the TDoA localization system.

The RT technique can model the propagation mechanism depending on the environment description details. In outdoor scenarios, like our TDoA localization environment, the main interactions are reflection and diffraction. Therefore we need to investigate the electromagnetic theory behind the RT formulation.

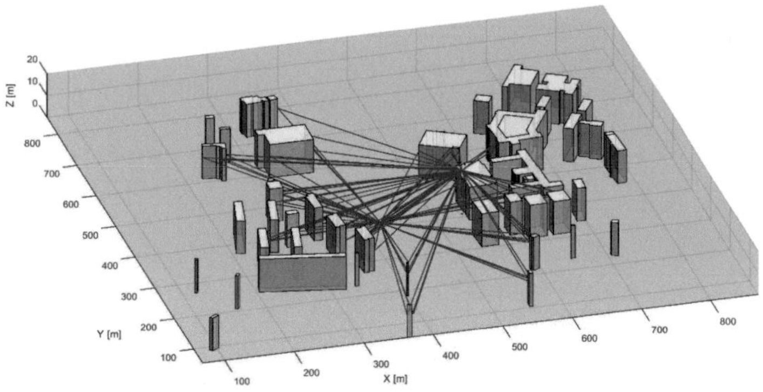


Figure 4.7.: RT Environment.

Starting with the reflection mechanism, consider the general case displayed in Figure 4.8. Asymptotic techniques are usually employed for electromagnetic analysis of large and complex scenarios.

Determination of ray paths from source to observation region is the first stage of computation, which is traditionally done using dyadic coefficients presented in [Balanis, 2012]. The dyadic approach is applied in electromagnetic problems to solve for multiple order interactions with scatterers in GTD or Physical Optics (PO).

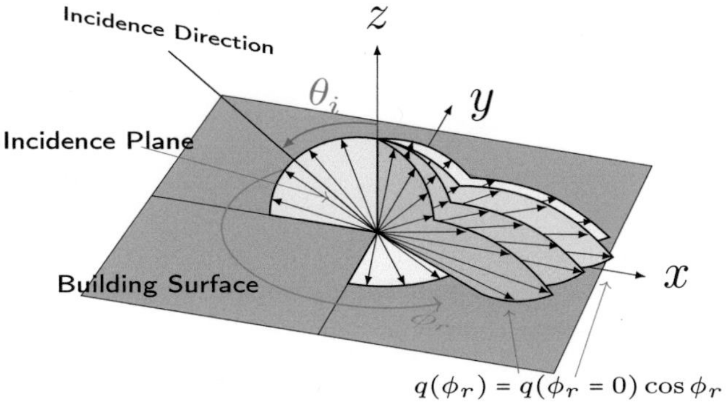


Figure 4.8.: The interaction between SC and obstacles, reflections follow different angles ($q(\phi_r)$).

Assuming that a SC hits $N_{EV}^k$ obstacles and each ray segment has a length of $s_l$, with the following structure:

$$\boldsymbol{E}_R^k = A_k(s_l, l = 0, 1, 2, \ldots, N_{EV}^k).\left[\prod_{l=min\{1,N_{EV}^k\}}^{N_{EV}^k} \underset{=l}{D}\right].\boldsymbol{E}_{TO}^k(\theta_T^k, \phi_T)^k.e^{-j\beta.s^k}, \qquad (4.2)$$

- $N_{EV}^k$ is the number of interactions of the $k^{th}$ ray, first assume only "bounces" in a reflection scheme;
- $s_l$ is the length of the $l^{th}$ segment of the $k^{th}$ ray;
- $s^k = \sum_{k=1}^{N_{EV}} s_l^k$ is the total length of all the rays that come out of the iteration;
- $\underset{=l}{D}$ is the dyadic, as defined by [Balanis, 2012], to decompose the field into perpendicular and parallel to the incidence plane;
- $A_k$ is the divergence factor.

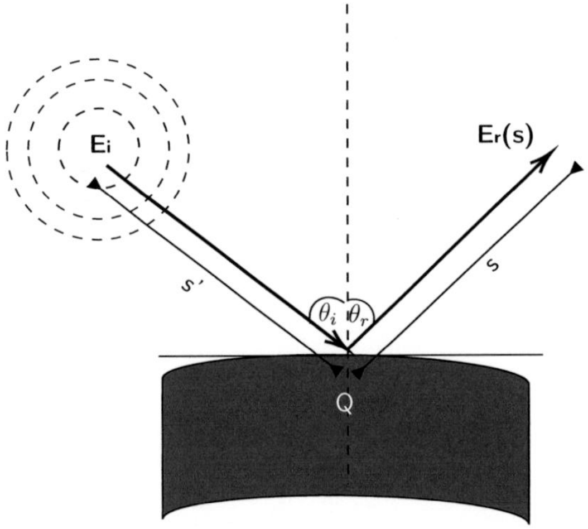


Figure 4.9.: RT Reflection, adapted from [Fuschini et al., 2008].

Figure 4.8 shows the interaction results in different angles. The ray that comes out of this interaction can use the reflection, diffraction, scattering, or refraction mechanism. In the case of reflection, the incidence ray and the reflected ray lay in the same plane ($q(\phi_r) = 0$).

Figure 4.9 shows a schematic view of a ray interaction in a curve surface, defining the reflection point as $Q_R$. The reflected ray has the following format:

$$\boldsymbol{E}^r(S) = \underbrace{\boldsymbol{E}^r(Q_R)}_{\substack{\text{Field at Ref}\\ \text{Point } (Q_r, s=0)}} \underbrace{\underline{\underline{\mathbf{R}}}(Q_R, \theta_i)}_{\substack{\text{Reflection Coeff}\\ \text{Dyadic}}} \underbrace{\sqrt{\frac{\rho_1 . \rho_2}{(\rho_1+s)(\rho_2+s)}}}_{\substack{\text{Spreading}\\ \text{Factor}}} \underbrace{e^{-j\beta s}}_{\substack{\text{Phase}\\ \text{Factor}}} \tag{4.3}$$

- $\boldsymbol{E}^r(Q_R)$ is the field at reference point $(Q_R, s = 0)$;
- $\underline{\underline{\boldsymbol{R}}}(Q_R, \theta_i)$ is the reflection coefficient in dyadic form, as defined in [Balanis, 2012];
- $\sqrt{\frac{\rho_1 . \rho_2}{(\rho_1+s)(\rho_2+s)}}$ represents the spatial attenuation factor, including divergence and spreading factor;
- $e^{-j\beta s}$ is the phase factor.

$$\underline{\underline{R}} = R_{//}\hat{e}^i_{//}\hat{e}^r_{//} + R_{\perp}\hat{e}^i_{\perp}\hat{e}^r_{\perp} \tag{4.4}$$

If the reflection in general curved surface is considered , the total reflected electric field ($TE$,$TM$ modes) in matrix notation is:

$$\boldsymbol{E}_r(S) = \boldsymbol{E}_r^{TE}(S) + \boldsymbol{E}_r^{TM}(S), \tag{4.5}$$

$$\boldsymbol{E}_r(S) = \begin{bmatrix} \Gamma_{TE} & 0 \\ 0 & \Gamma_{TM} \end{bmatrix} \begin{bmatrix} \boldsymbol{E}_i^{TE}(Q_R) \\ \boldsymbol{E}_i^{TM}(Q_R) \end{bmatrix} \sqrt{\frac{\rho_1^r \rho_2^r}{(\rho_1^r + s)(\rho_2^r + s)}} . e^{-j\beta s} \tag{4.6}$$

$\rho_1^r \rho_2^r$ are the main curvature radii of the reflected electric field for a generic wave.

In our case, the majority of reflections in an outdoor scenario are caused by buildings walls. Thus reflection effect can be simplified to a plain surface interaction, in this situation Equation (4.6) is simplified to Equation (4.7), assuming $\rho_1^r = \rho_2^r = s'$.

$$\boldsymbol{E}_r(S) = \begin{bmatrix} \Gamma_{TE} & 0 \\ 0 & \Gamma_{TM} \end{bmatrix} \cdot \begin{bmatrix} \boldsymbol{E}^0_{TE} \\ \boldsymbol{E}^0_{TM} \end{bmatrix} \frac{e^{-j\beta s}}{s'} \sqrt{\frac{(s')^2}{(s+s')^2}} = \boldsymbol{E}^0 \underline{\underline{R}} \frac{e^{-j\beta(s+s')}}{s+s'} \tag{4.7}$$

After analyzing the reflection interaction, we now investigate the diffraction propagation mechanisms. In electromagnetic theory, the UTD is a high-frequency method for solving scattering problems from electrically small discontinuities, which is a typical phenomenon when the rays hit the edges of the obstacles.

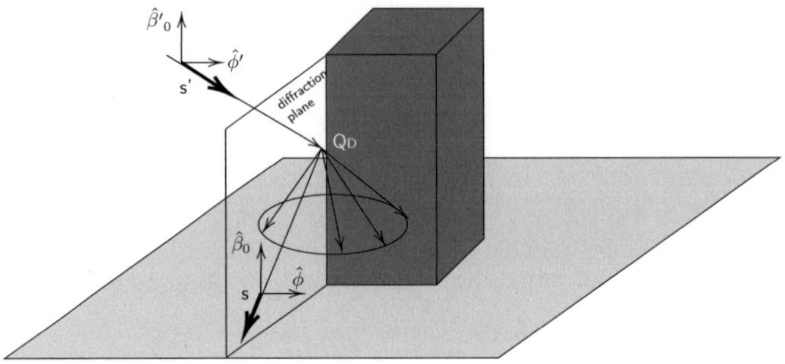


Figure 4.10.: Hybrid approach: SBR and Image Theory.

The near-field electromagnetic fields are assumed quasi-optical. The diffraction coefficients are determined and used to calculate the field strength and phase for each direction away from the diffracting point. These fields are then added to the incident and reflected ones to obtain a total solution.

Figure 4.10 shows the diffraction effect schematically when the ray hits an obstacle edge. The propagation mechanism is the diffraction, modeled as:

$$\begin{bmatrix} E^d_{\beta_0} \\ E^d_{\phi} \end{bmatrix} = \begin{bmatrix} D_s & 0 \\ 0 & D_h \end{bmatrix} \begin{bmatrix} E^i_{\beta_0} \\ E^i_{\phi} \end{bmatrix} A(s, \rho^d) e^{-j\beta s} \tag{4.8}$$

the term $A(s, \rho^d)e^{-j\beta s}$ being the spreading factor for diffraction. Using the concept of "unfolded" fields as discussed in [Fuschini et al., 2008], the diffracted rays must satisfy, such that: $\beta_i = \beta_d$

$$A(s, \rho^d) = \sqrt{\frac{\rho^d}{s.(\rho^d + s)}} \tag{4.9}$$

Figure 4.11 shows reflection and diffraction as isolated entities, simulated with RT. Although depending on the RT definition, one ray can suffer several effects in series, Figure 4.12 shows the double interaction, where a SC faced a reflection followed by diffraction.

Starting with the reflection effect, the electric field at point $Q_R$ is:

$$\boldsymbol{E}(Q_R) = \boldsymbol{E}^0 \frac{e^{-j\beta s"}}{s"} \tag{4.10}$$

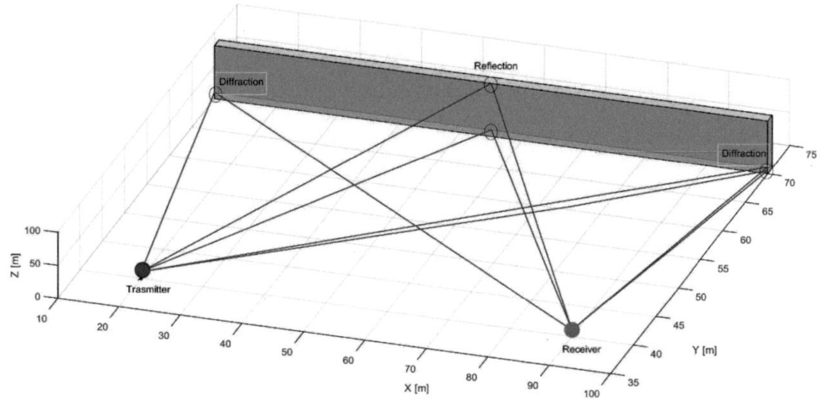


Figure 4.11.: Reflection and Diffraction Interaction.

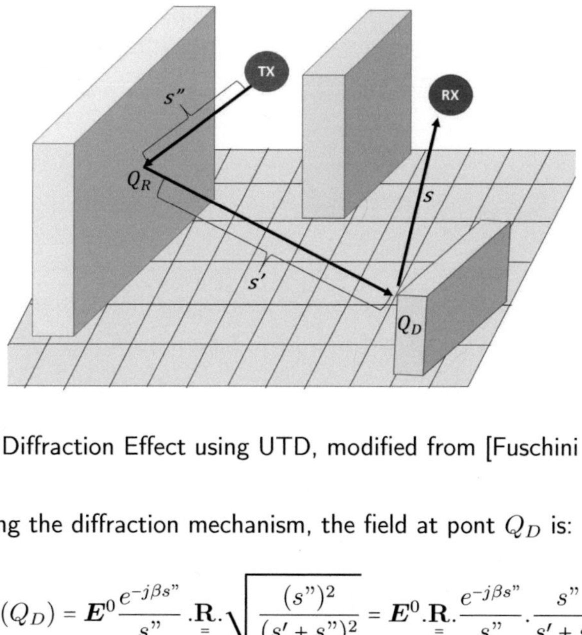


Figure 4.12.: Diffraction Effect using UTD, modified from [Fuschini et al., 2008].

Now, including the diffraction mechanism, the field at pont $Q_D$ is:

$$\boldsymbol{E}(Q_D) = \underbrace{\boldsymbol{E}^0 \frac{e^{-j\beta s"}}{s"}}_{\boldsymbol{E}(Q_R)} . \underline{\underline{\mathbf{R}}} . \sqrt{\frac{(s")^2}{(s' + s")^2}} = \boldsymbol{E}^0 . \underline{\underline{\mathbf{R}}} . \frac{e^{-j\beta s"}}{s"} . \frac{s"}{s' + s"} \tag{4.11}$$

and the combined effect can be structured as follows:

$$\begin{aligned}\boldsymbol{E}(Rx) &= \boldsymbol{E}(Q_D).\underline{\underline{\mathbf{D}}}.\sqrt{\frac{(s' + s")}{s\left[s + (s' + s")\right]}}e^{-j\beta s} \\ &= \boldsymbol{E}^0.\underline{\underline{\mathbf{R}}}\underline{\underline{\mathbf{D}}}\frac{1}{s"}\frac{s"}{s' + s"}\sqrt{\frac{(s' + s")}{s\left[s + (s' + s")\right]}}e^{-j\beta(s+s'+s")} \\ &= \boldsymbol{E}^0\underline{\underline{\mathbf{R}}}\underline{\underline{\mathbf{D}}}\frac{1}{\sqrt{s(s' + s")(s + s' + s")}}.e^{-j\beta(s+s'+s")}\end{aligned} \tag{4.12}$$

To put it concisely, RT models the multipath in a given simulation domain and shows how the obstacle affects the CIR. The desired number of rays and interactions is predefined in the simulation; however, it can be boundless ideally because there are multiple possible paths in an urban environment.

The scenario elements, including buildings, trees, cars, pedestrians, and other structures, are an approximation of the real world. Therefore, the level of details and spatial discretization directly impact the computational requirements to perform the simulation.

## 4.4. Limitations of RT Tools in CIR Estimations

As a numerical method for representing the ray propagation, the RT itself has limitations related to the theoretical foundations used. For example, both GO and UTD assume that the dimension of the obstacles is greater than the wavelength of the ray.

The GO method is based on the assumption that electromagnetic waves travel in straight lines in the form of rays and that they are reflected from the illuminated surfaces of an object. The main limitation of GO is that it ignores the wave properties related to diffraction, interference, and polarization. It is not a problem for large-scale interactions, where such effects may be negligible or separately taken into account. The UTD can add the scattering effects caused by the electrically small objects or discontinuities in more than one dimension at the same point.

Like most of the applications of high-frequency methods, these approaches deal with the point-to-point construction of rays and do not consider the field behavior in a complete domain. As a result, they cannot accurately handle inhomogeneous media and variations in the refractive index of the troposphere in propagation problems.

Modern studies include the DMC in RT simulations. However, this approximation is only valid when the environment is described with high fidelity details, including windows, vegetation, and roof description, as presented in [Fuschini et al., 2015].

The position fingerprints method proposed in our investigations relies on the capacity to extract the CIR multipath characteristics of each point of the scenario. Therefore the following limitation aspects of RT must be taking into account for radio propagation prediction:

**High-Frequency Performance:** RT is a high-frequency asymptotic solution of electromagnetic Maxwell's equations, which means that it produces accurate results in high frequency. However, there are differences between the prediction, and the CIR increases in lower frequency. Thus, the fingerprints, for the same scenario, presents better results for higher frequency simulation.

**Dependency of the Scenario Description:** The CIR estimation accuracy depends on the obstacle details in the scenario modeling. It is usual to take building descriptions from open-source databases [Bennett, 2010] but are usually inaccurate. It might be possible to use LIDAR building description [Bedford et al., 2019], making RT simulation more accurate, but in most cases, this is not a feasible option.

**Propagation Mechanism is Oversimplified:** RT simulations are usually performed by assuming that the ray does not penetrate in the materials, and only reflection and diffraction occur. There are much more complex interactions involving multiple reflections, diffractions, and scattering mechanisms in a real-world scenario. In a large-scale observation, these multiple interactions tend to fade, but it might be possible the transmission through buildings.

**Obstacle Roughness:** The obstacle roughness depends on the frequency adopted in the simulation. This value can define if a reflection is specular or diffuse, at frequencies higher than 10 GHz, the roughness of the building becomes significant and need to be included in the RT model to reproduce the reality.

**Emitter and TDoA Height:** The contribution of the lateral component of the electromagnetic wave depends on the height of the emitter and the position of the sensors. If the emitter is well below average building height, lateral propagation is more important; if it is located above the average building height, vertical over rooftop propagation is significant, [Fuschini et al., 2015].

**Bandwidth and Material Parameters:** It is possible to use RT for broadband signals; in this case, the simulation is performed, setting the excitation signal for each frequency in the desired bandwidth. The problem is to define the material parameters in each frequency properly. The SC simulated in RT are impulse responses for a Dirac in time-domain, therefore with infinite bandwidth. It is challenging to include dispersive material characteristics, and dielectric properties [Fuschini et al., 2015].

In RT simulation, the size, the location, and the number of reflectors, transmitters, and absorbers are specified *a priori*. Thus, the accuracy depends on the output level of precision of the parameters defined.

In an outdoor scenario, it is challenging to define material and obstacle details with high fidelity; consequently, the simulation is a mere approximation of the real-world situation. The multipath fingerprints can use RT. Nevertheless, it is expected that some grade of mismatching exists between the information extracted from the signal and the output of the RT simulation.

## 4.5. Multipath Fingerprints with RT

This section explains the use of RT to build the multipath fingerprints that form the core of the proposed method. The advantage of deterministic propagation tools like RT is that they take into account the effects of all obstacles within the scenario, including conductivity and dielectric constant at the frequency range of operation under analysis.

RT simulation provides the "site-specific" channel impulse response. Therefore, if the sensor location is defined, the software extracts all multipath information for each point of the scene. Figure 4.13 shows an example of RT mapping, we can see the rays arriving in the Sensor 2, and the final CIR composed of a combination of SC rays for each position in the simulation domain.

We proposed establishing a one-to-one relationship between the signal received at the base station and the emitter position in this work.

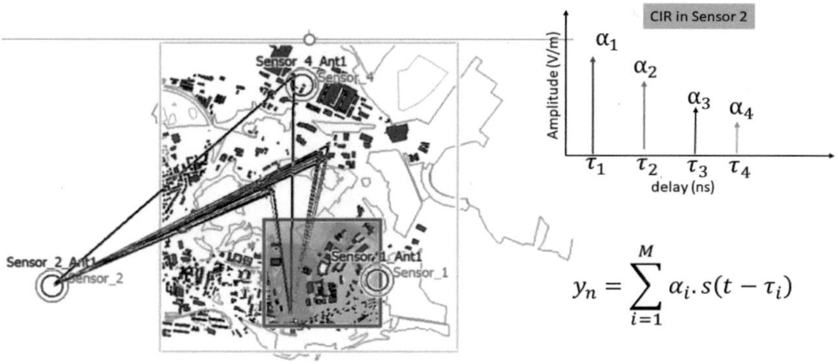


Figure 4.13.: Multipath Fingerprints with RT.

The fingerprints are extracted from the received signals and serve as a unique identifier of the emitter location. Therefore, the localization is reduced to a pattern recognition problem, where the position is determined by matching the extracted fingerprints with the fingerprint database.

The RT provides not only amplitude and delay patterns $(\alpha_i, \tau_i)$ but also gives information related to the reflection and angular points of each ray.

The actual RT implementation, mainly in an outdoor scenario, adopts a certain level of approximation, either in the obstacle description or in the material properties. Therefore, the *cubic* buildings used in the simulations are somehow a rough approximation of the highly complex surface that defines the obstacles.

As reported in Section 4.2.4 and Section 4.3, the scenario objects are simplified, with general information of the electromagnetic material properties. Moreover, if there are several different materials in the environment, only the major ones are included in the simulation.

RT is a deterministic tool, yet in an outdoor scenario, it uses a statistical approximation to characterize the radio wave propagation in the respective urban scenario.

Figure 4.14 shows an outdoor scenario split into small squares displaying a grid of possible emitter positions. Assume we have four TDoA sensors; there is a particular combination of the CIR arriving in each sensor.

Referring to the collecting region concept, as explained in Section 4.2, the central point in the simulation approximates all points inside the square grid (or cube in 3D). Therefore, the collection region defines a certain degree of position approximation. As far as those points are in the same grid, their CIR fingerprints are the same.

The modern RT approaches in [Charbonnier et al., 2017] and [Steinböck et al., 2016] have included the so-called hybrid methods to use stochastic and deterministic schemes in the propagation simulation.

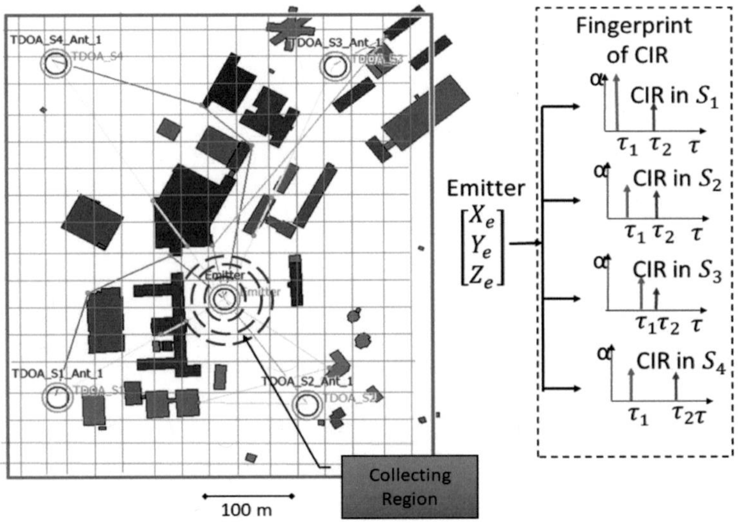


Figure 4.14.: Multipath Fingerprints with RT.

The connection of RT with TDoA localization has two approaches. One of them is using the pure multipath information directly in the SIMO FIR CIR estimation, as shown in Chapter 3. Also, the CBB received in each sensor needs to be decomposed to extract the $\alpha, \tau$ pattern.

Another approach is to produce a synthetic cross-correlation function, where the multipath is hidden in other maximums of the VCC operation, explained in Section 3.5. Moreover, other peaks are not used to extract the time differences once only the lag that represents the highest peak is considered.

### 4.5.1. Multipath Fingerprints SIMO approach with RT

Multipath fingerprints can be built by taking the RT output directly. Then, defining each sensor position, the CIR information can be extracted for each point of the scene.

We have only used the delay and amplitude parameters in this method; however, RT output gives all reflection and diffraction positions and the complete geometric path of each valid ray. In addition, the RT output also provides information about the propagation mechanisms that are affecting the rays.

The ray-tracing scenario is decomposed into walls, edges, and the "view tree." It represents the interactions between emitter and sensor, which forms the basis for the visibility matrix, where each interaction is considered a layer in a multilayer scheme. Therefore, RT provides information about each ray path, describing the edges and the walls touched by the rays in the emitter-sensor path.

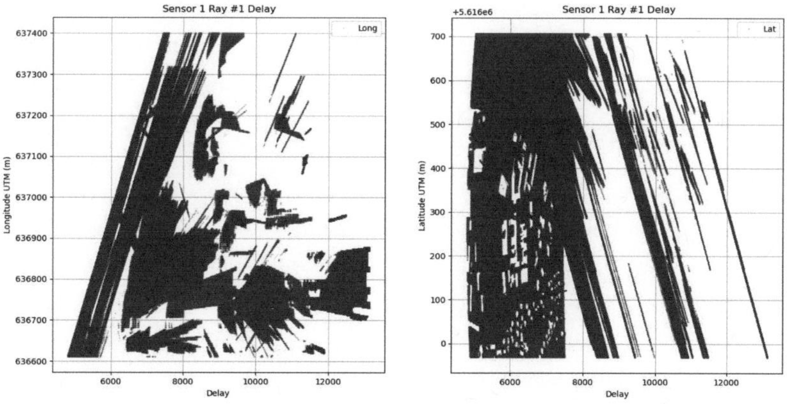


Figure 4.15.: Delay Multipath Fingerprints of Ray 1 in TDoA Sensor 1.

The fingerprint idea can be explained in Figure 4.15 (2D) and in the Figure 4.16 (3D). They both show the multipath fingerprints produced by the first ray delay ($\tau_1$) that arrives in Sensor 2. The left one shows the fingerprints for Longitude in UTM, and the right one shows the latitude patterns.

Figure 4.16 shows the same Delay Fingerprints for Ray 1 in a 3D view, in which we can observe the shape of the buildings in the empty spaces where there is no ray propagation

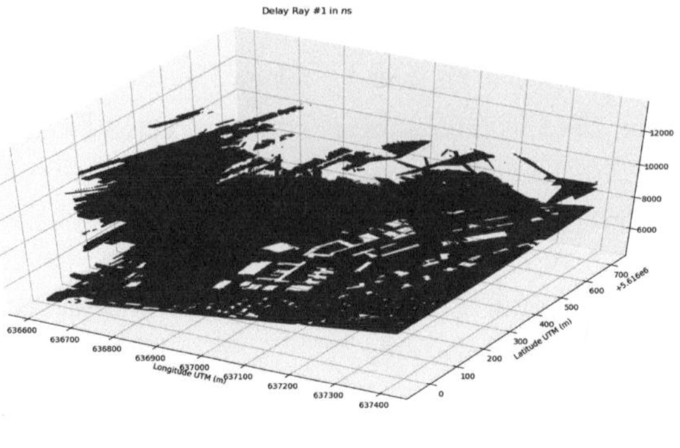


Figure 4.16.: Multipath Fingerprints with RT in 3D.

from internal parts of the building structures. Each point has a defined delay in $ns$; each ray amplitude and delay is a characteristic of the position, as much feature we consider as much uniqueness the fingerprints become.

Then each latitude-longitude defines a Point that has a value to describe the delay of the first ray. It is possible to have more than one point with the same delay values. When we increase the number of rays taken into consideration, the features for each point start to become unique.

By considering the first rays only, there is no one-to-one relation between position and feature. However, if we repeat the analysis by adding more features, like the ray amplitude, given in Figure 4.15, we add a new classification layer. For example, if we consider 20 rays, it will have a classification dimension of 40, as each ray has the $\alpha, \tau$ value to compose the multidimensional classification engine.

### 4.5.2. Multipath Fingerprints VCC approach with RT

Remembering the aspects explained in Section 2.4.3 and Section 3.5, a multipath pattern can also be extracted from the cross-correlation operation.

VCC gives a better resolution than MUSIC and the usual cross-correlation methods. The TDoA extraction only takes the lag and the maximum values of the cross–correlation operation.

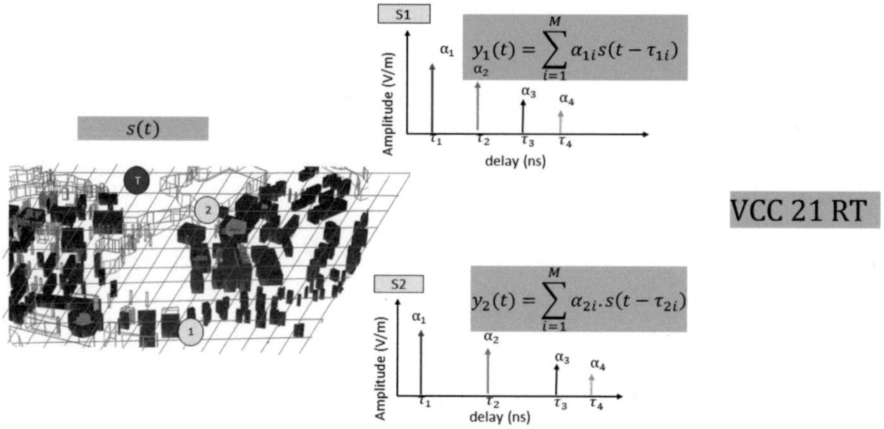


Figure 4.17.: Multipath Fingerprints with RT.

The other maxima that carry the multipath information are not used. Hence, the idea is to use this step in the TDoA positioning processing to enhance the estimation with the multipath information.

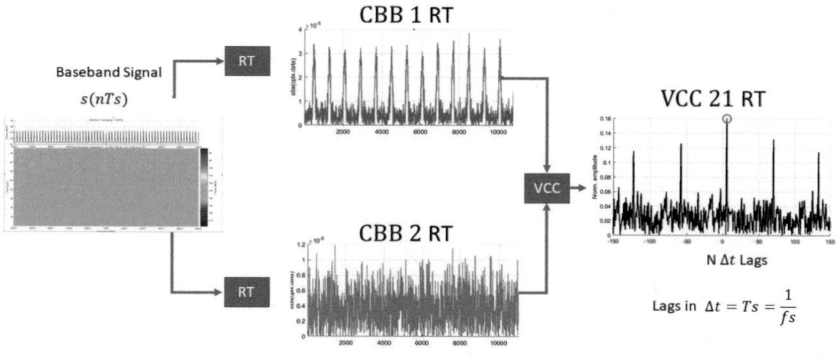


Figure 4.18.: Generating Multipath Fingerprints with RT.

Figure 4.17 shows the signal arriving in each sensor ($S1$ and $S2$), where the CIR information (provided by RT) is used to produce two synthetic CBB signals.

It is necessary to explain that this operation does not affect the blind nature of the localization because the $s(t)$ signal used to produce the synthetic CBB is only a signal with the same structure ($BW$) from the target signal.

The two CBB signals produced by employing RT simulations and a sample signal $s(t)$ are used to produce a synthetic signal. Figure 4.18 shows a VCC generated by using the RT output.

The final multipath fingerprints are assigned for each position as a set of VCC, from the pairs of TDoA sensors.

## 4.6. Concluding Remarks

In this chapter, the concept of ray tracing, along with its techniques, was explained in detail. The different procedures employed in ray tracings such as SRB, Image Method, and Hybrid Model are introduced and explained.

The main aspects of RT algorithms were investigated. There are several advanced simulation tools, but we did not focus on a specific one. Therefore, we have adopted and simulated some standard algorithms, showing some electromagnetic aspects of the GTD and UTD inside the RT itself.

The visibility tree and the collecting region were discussed in detail, determining the relevant rays in the simulation were also described. Finally, we have demonstrated the trade-off between the collecting region size and how it impacts the grid size in the fingerprints.

The issues about outdoor environments were also reported, where we simulated and analyzed the impact of obstacles and discussed the simplification methods applied to them. Finally, we have shown the use of the RT results to build multipath fingerprints, either using the CIR directly or using a synthetic VCC scheme.

The approach presented here plays a fundamental role in the proposed method; it is applied together with the concepts introduced in Chapter 2 and Chapter 3 for multipath exploitation. The idea is to bind the ray-tracing with the CIR estimation and add a machine learning engine to enhance the localization performance.

# 5. Machine Learning in Localization Problems

In conventional radio-frequency localization techniques, we use triangulation rules in Euclidean geometry to infer the emitter's position. As previously mentioned, it is also possible to generate a machine learning approach for exploiting the multipath information from the CIR in each sensor, as discussed in Chapter 3.

Consider an urban outdoor scenario, where sensor position and general building description are fed to a ray-tracing simulation for building a multipath fingerprint dataset. This multipath information exploits the site-specific characteristics and assists us in building a machine learning engine to improve emitter estimation. Various machine learning algorithms can be used to localize the emitter's position. In this work, we use a well-known machine learning technique called *Random Forest*.

This chapter addresses the general formulation of machine learning techniques for positioning problems and ray-tracing simulations to produce the dataset fingerprints. We divide this chapter into three parts.

In the first part, we present a general discussion on using machine learning for localization. In the second part, we focus on the implementation of the Random Forest technique. In the third part of this chapter, we discuss the necessary adjustments and considerations to build a machine learning framework using mixtures data produced by ray-tracing simulations and the real measurements from reference emitters. This chapter describes how randomization characteristics introduced into the decision trees can decrease the prediction error.

## 5.1. Motivation and Related Work

There are various works presented in the recent literature that use machine learning methods for object localization. Authors in [Tran and Nguyen, 2008] propose a variation of support vector machine (SVM) for localizing the sensor in a network based on connectivity information. The proposed method exhibits fast localization by performing computations in a distributed manner and by efficiently using computation and communication resources.

The standard localization techniques use the minimal error between two neighbor nodes, typically employing Euclidean distance. We focused herein on the TDoA approach, but the pair-wise distance could also use Received Signal Strength Indication (RSSI) [Whitehouse, 2002], ToA, or AoA. Many techniques that employ concepts from machine learning have been proposed [Brunato and Battiti, 2005, Van Nguyen et al., 2015]. The essential idea of these methods is that it is possible to exploit the topology information implicit in sensor locations in the construction of possibly non-Euclidean function spaces that are useful for estimating unknown emitter locations. The multipath fingerprints procedures explained in Section 4.5 can be used as the training data for a machine learning procedure. Therefore, a prediction model can be built to enhance the overall localization performance.

In [Chen, 2012], the authors propose a hybrid approach by combining ToA and AoA measurements with an Artificial Neural Network (ANN) for enhancing the accuracy of mobile station location prediction. The ANN overcomes the problem of nonlinearities and estimates precise locations in different NLOS environments.

The authors in [Syldatk, 2012] investigate the effects of calibration of ground sensor networks on the performance of detection, classification, and tracking objects of interest. The authors use the expectation-maximization (EM) algorithm. The EM algorithm uses the positional data of reference emitters to reduce the unknown calibration parameters (biases). Besides, this algorithm provides a robust and accurate estimation for calibrating the sensor networks using opportunistic targets.

In [Ziedan, 2012], authors address the error caused by non-line-of-sight signals in the global navigation satellite system (GNSS). The authors employ a machine learning algorithm that uses multipath signals reflected off different surfaces and is propagated to the receiver afterward.

In [Guan et al., 2013], the authors present an Elman neural network to estimate and mitigate NLOS error in TDOA positioning. The paper presents improvements in simulations to enhance position estimation.

[Alsheikh et al., 2014] presents an in-depth literature review for using machine learning in wireless sensor networks (WSNs). The authors compare various machine learning algorithms by highlighting the advantages and disadvantages in each case. In addition, the study also provides a general guideline to apply machine learning in network localization problems.

The authors in [Mahfouz et al., 2014] outline a method for target tracking by combining machine learning with Kalman filter to estimate instantaneous positions of a moving emitter. The authors also develop radio fingerprints based on received signal strength indicators (RSSI). Then, they use this dataset to train the machine learning algorithms for estimating the target's position. In addition, the authors apply the Kalman filter on predictions of the target's position.

The authors in [Van Nguyen et al., 2015] present an algorithm for enhancing the accuracy of time-of-arrival localization by identifying and mitigating NLOS signals with relevance vector machine (RVM). The proposed algorithm is a two-step iterative algorithm and combines stochastic localization networks with cooperative localization methods.

In [Zhao et al., 2014], the researchers present a technique to improve the localization accuracy in multipath environments. The performance improvement is accomplished by introducing a localization approach based on the k-nearest neighbor (kNN) algorithm. The researchers use the distance from the reader's antenna to a target tag as a weighting factor.

In [Klautau et al., 2018], authors propose a deep learning (DL) algorithm for performing localization by connecting a vehicle traffic simulator with a ray-tracing in 5G scenarios. The authors also conduct multiple experiments using deep learning for classification, regression, and reinforcement learning problems.

In [Bibb et al., 2016], researchers investigate the radio frequency emitter localization in urban environments using machine learning methods. The authors use both the classification techniques and the regression schemes for this purpose.

In [Quan, 2017], the authors develop a multipath detection model based on machine learning methods and present an effort to find a multipath signature in the received time series of satellite measurement signals from GNSS (GPS, GLONASS, Galileo, BDS, and QZSS).

The researchers in [Li et al., 2017] used a kernel-based machine learning localization scheme applied to NLOS situations, where the TDoA proposed solution is insensitive to synchronization errors.

The authors in [Suresh et al., 2017] propose a technique that uses K-means clustering and support vector machines (SVM) to exploit the data patterns produced by spatially distributed sensors. The authors also emphasize the importance of data pre-processing to enable accurate data association from different sensors for localization.

In [Navabi et al., 2018], the researchers investigate the viability of using machine learning techniques for estimating user-channel features at a large-array base station (BS). The authors propose using a pilot signal broadcast for extracting propagation channel angle of arrival (AoA), received signal strength, and relative path delay. The authors use this information for training a neural network to predict the angle-of-departure (AoD) of the dominant propagation paths in the user channels. In addition, the authors, based on standard statistical performance, prove that the data-driven methods can predict unobserved channel features from observed ones.

The investigations presented here used as a general rule the idea of machine learning applied to the location of an emitter. Most of them utilized the RSSI technique to make a fingerprint of the power distribution. The Fingerprint part of our method is based on this method, but instead of using RSSI, we have included the TDoA approach as a position feature. In our concept, we have developed the idea of using synthetic and real measures to make the datasets. In this manner, the suggested machine learning techniques used a mix of RT simulations and actual measurements to produce the position fingerprints.

As presented, several localization applications use machine learning to either enhance or improve the performance, specifically when the multipath effects deteriorate the overall results. The research outcomes presented here form a basis for developing the proposed framework. In this work, we use machine learning, especially in classification setup, and add the ray-tracing fingerprints from the channel multipath information available in the cross-correlation of the received signals.

## 5.2. General Framework of Machine Learning

In several fields of science, it is necessary to predict a phenomenon based on previous measurements or observations. For instance, we may have datasets of recorded signals and known position emitters in localization problems that model some previous environment or phenomenon. These datasets are very useful in the development of machine learning techniques for solving the localization problem.

In this work, we use past recordings to extract a model that can efficiently "predict" the target's position based on the instantaneous observations. Moreover, it is also possible to set up the localization systems to learn from the sensor's measurement data and estimate the unknown target's position.

The authors in [Mitchell et al., 1997] discuss machine learning algorithms that can learn from data without being explicitly programmed. In other words, a computer program is supposed to learn from data when a defined task receives more and more data to reinforce the learning process. In this sense, machine learning provides algorithms that solve classification or regression tasks. It is accomplished by building a model to predict new states using the structure of the data [Breiman, 1996]. Machine learning models are used in three different settings: supervised learning, unsupervised learning, and reinforcement learning.

In this work, we investigate the localization problem in terms of reinforcement learning. We assume a reinforcement procedure with real measurements from reference points. In the machine learning approach, we use measurements to train a model to refine and make it a generic model that can cope with new data.

The goal is to predict the position based on observed predictor variables (reference emitters).

For generality, let us assume that the TDoA system has $N$ sensors $\{S_1, S_2, \ldots, S_N\}$ deployed in an outdoor environment. To keep the analysis simple, we consider only two dimensions. Please note that the techniques apply to the three-dimensional models as well.

There are $k$ reference emitters whose locations are known, i.e., $\{R_1, R_2, \ldots, R_k\}$. The process of mapping the input-output relation between the features and the labels is the key idea of model training. If we define $M$ as the total number of different locations, the learning-based algorithm aims to estimate the position using every possible location, i.e., $\{R_{k+1}, R_{k+2}, \ldots, R_M\}$, which has a specific CIR extracted by each TDoA sensor.

In Figure 5.1, we present a possible machine learning framework for the implementa-

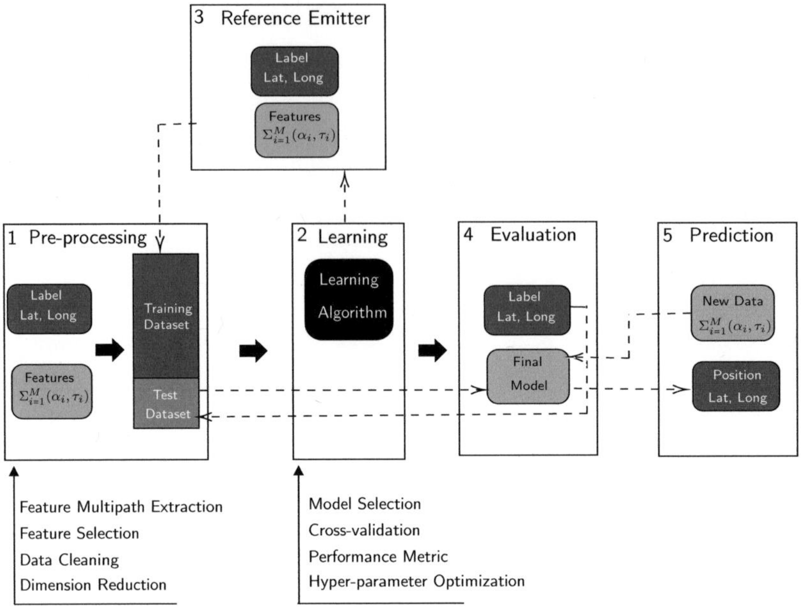


Figure 5.1.: Framework of Machine Learning Implementation.

tion of source localization. The figure displays the general building blocks of the considered technique.

- In Step 1, we pre-process the dataset. We consider that the emitter coordinates are labels, and the multipath information represents features. The model uses the CIR estimation or the VCC peaks of the complex digital base-band signals that arrive in the sensors. Also, we perform the following activities in this step:

  - We define feature relevancy, i.e., which features are more relevant based on [Landmann, 2008]. We define five peaks in the VCC or the CIR results.

  - We split the dataset into two parts, the "training dataset" used to build the model and the "test dataset" to evaluate the trained model.

  - We perform "data-cleaning," i.e., removing noise, normalizing the features, and removing the outliers;

- We perform dimensionality reduction, i.e., we decrease the number of features in the dataset, as some of the features may present high correlation among them. The process of dimensionality reduction reduces the feature space, consequently reducing storage and computational requirements.

- The machine learning model is built, in Step 2, after comparing different algorithms to train and select the best model. It is necessary to decide upon a metric to measure performance based on classification accuracy (the proportion of correctly classified instances).

- In our approach, first, the model is trained with a simulation dataset, built using only RT; after that, the data is "enhanced" with real measurements from the environment. In Step 3, we add the features and labels of the reference emitters in the first dataset. We use it to train and refine the model, achieving a better performance in tuning the learning algorithm.

- In Step 4, we evaluate the model using the features from the "test" dataset and try to predict the emitter's position (labels). The central idea is to evaluate models and predict unseen data instances, to estimate how well it determines the generalization error.

- In Step 5, we use this model to predict new unseen data, using the enhanced model achieved.

With this general framework in mind, the next step is to describe how the machine learning approach can use the ray-tracing simulation for enhancing TDoA localization systems. Thus it is necessary to demonstrate how the localization fingerprint in this thesis was organized in machine learning.

Assuming that the unknown emitter $P_e$ is in the position defined as $(x_e, y_e)$ on a 2-D field, we know with some precision level the exact position of each "$i-th$" sensor in the localization system ( $S_i$ ). We divided the scenario using the approximation rule in a grid of possible points where we have the emitter position.

It is crucial to evoke the concept of "collecting region" introduced in Section 4.2 because the grid size assigns all points in this grid as a unique position. If the collecting region has the same order of grid size, we can assign at the same time the collections of place in the grid for machine learning and RT.

Some estimations may be in the same region, indicating a clustering on the features or some incapability of classification. Every grid or "small region" is assumed as one class in a classification task, as shown in Figure 5.2, where the classes are $R_1, R_2, R_3, R_4$.

Based on these memberships, we can localize $P_e$ using classification procedure, as a member of class of $R_1$, of $R_2$, of $R_3$ and of $R_4$; then $P_e$ must be in the inter-sectional area $R_1 \cap R_2 \cap R_3 \cap R_4$.

It is also feasible to solve the localization problem using a regression approach, as introduced by [Pan et al., 2006] to infer the emitter position based on the signal measured in each sensor.

We have directed our investigation toward TDOA systems, but this approach can also use signal strengths (RSSI). The regression tool can map the signal features directly into an emitter position estimation.

The execution of this approach is usually more straightforward than the localization techniques based on geometry. [Nguyen et al., 2005] showed a scheme to use the RSS and mapping the values as input to the learning or TDOA measurements from the sensor systems. Multipath components from the NLOS condition corrupt the performance in urban and suburban scenarios; the correlation between the TDOA and physical location space is usually non-linear.

It is also challenging to know in advance, given a sensor structure, the exact features that uniquely identify the emitter location. There is a strategy introduced by [Schölkopf et al., 2002] for learning the correlations between position and signal strength using kernel methods for statistical classification and regression.

[Louppe, 2014] showed that a set of possible position in the outdoor scenario, defined by $P_e$, can produce an algorithm ($\boldsymbol{A}$ based on the position-multipath fingerprints relationship). The model produced $\boldsymbol{A}(\theta, P_e)$, denoted $\varphi_{\boldsymbol{L},\theta}$, where $\theta$ corresponds to the hyper-parameters controlling the execution of $\boldsymbol{A}$.

With this general framework in mind, the next step describes how the machine learning approach can use the ray-tracing simulation and enhance the TDOA localization systems. For this reason, it is essential to show how the localization fingerprint proposed in this thesis was orchestrated in machine learning.

The positions with less than $5$ paths represent a problem for the formulation because

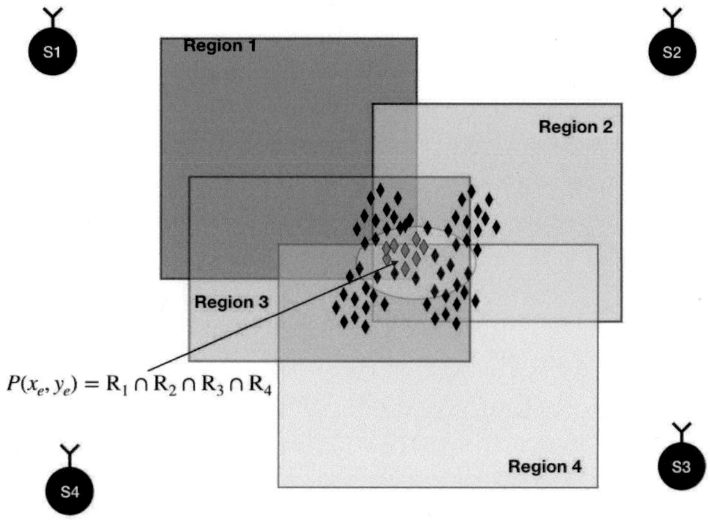


Figure 5.2.: Assuming that the geographic region in the scenario subdivided in a set of classes, the emitter locations can be estimated based on the intersection.

it is necessary to assign them with 0s to make them with the same length as the other network inputs. The *zero-force* procedures cause biases in the estimation. Consequently, the high values for the weights in the training procedure produce an imbalance in the hyper-parameters used in linear regression [Neter et al., 1989].

## 5.3. Multipath Fingerprint Datasets

The multipath fingerprints datasets used is a mixture of real measurements based on CIR extraction or VCC, as shown in Chapter 3; and the RT simulations as introduced in Chapter 4.

Assuming that each point in the grid of the outdoor scenario is a class of a region discretized in a $2$ meters grid, a $2 \times 2$ km will generate $1 \times 10^6$, different classes. The $N$ features can generate huge datasets used to train the model to extract the most relevant features of the scenario.

Each of these datasets gives a mapping between the position (lat, long) and features (multipath information). The union of the data from measurements and the simulation is done using a machine learning approach, defined as "data-augmentation" [Yu et al., 2017].

Sometimes it is not easy to have real measurements. In radio-frequency fingerprints, an emitter would be necessary for each grid position, which is not feasible. That is why data augmentation is necessary a concatenating procedure to assign one or more features to the same class (point in the grid space).

As described by [Sinha and Hwang, 2020], a machine learning performance depends on large amounts of data to train the models.

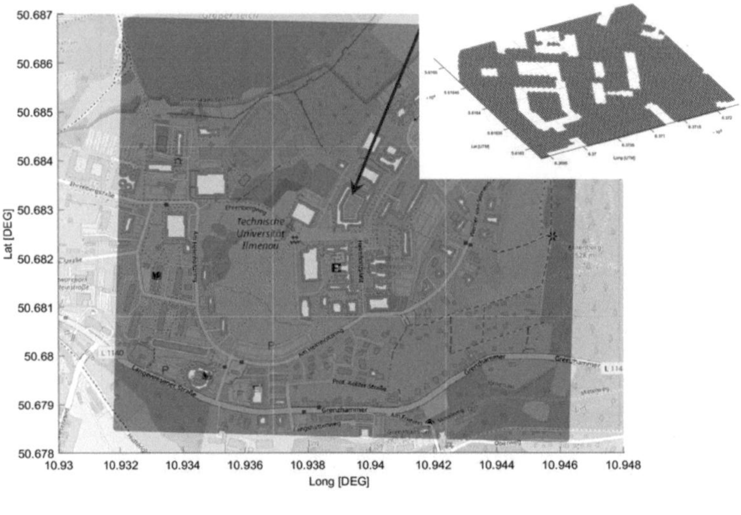


Figure 5.3.: Multipath Dataset as a Grid using RT.

As a real dataset based on measurements is usually unfeasible, we propose an improved data augmentation technique. Therefore, we concatenate RT simulated data with some actual measurements to enhance the machine learning models. A similar procedure was applied to received signal strength indication (RSS) values for fingerprint indoor positioning systems [Schumann et al., 2017].

Figure 5.3 shows how the dataset based on multipath fingerprints is created. We can see that the area was discretized in a grid, where each point represents a small region where the emitter could be. It is necessary to recall the concept of RT collecting region, introduced in Section 4.2.1 because the grid size is related to the region assumed with the same features.

In our case, the dataset is amplified with more features assigned to the same position. The overall effect of this approach improves the learning models, either for classification or regression. The hyper-parameters and the model are tuned with more values giving better prediction performance.

## 5.4. Decision Tree in Machine Learning Models

There are several methods to implement machine learning models. As indicated in Section 5.2 , most modern machine learning implementations use decision trees as a base algorithm; here, we will focus on the forests of randomized trees [Ho, 1998] designated as **Random Forest**.

The Random Forest machine learning algorithm has its name "forest" related to building an ensemble of decision trees grown from a randomized variant of the tree induction algorithm [Alsheikh et al., 2014].

Decision trees are suitable for modeling the non-parametric structure because they can deal with arbitrarily and complex relations between inputs and outputs, without any *a priori* assumption. Moreover, they can process heterogeneous data using selecting the essential features selection approach, making them robust to noisy variables.

In multipath fingerprint measurements, noise leads to wrong information in delay, amplitude, or the peaks of VCC function. The decision trees are also robust to outliers and errors in label information. Consequently, it is possible, in our localization scenario, to map the decision trees to sub-regions.

From a geometric point of view, the principle of the structured tree consists of approximating the partition of the Bayes model by recursively splitting the input space $\boldsymbol{X}$ into sub-spaces and then assigning constant prediction values $\hat{y} \in \boldsymbol{Y}$ to all objects $\mathbf{x}$ within each subspace. The Random Forest can be implemented as a **regression** or a **classification** engine. There are some studies using regressions in localization with CIR fingerprints in [de Sousa and Thomä, 2018b].

It is challenging to use regression formulation [Sohn et al., 2007], because multipath dataset structure is highly non-linear. The classification uses multipath as a feature matching in decision trees. A classification procedure of one of the random trees is indicated in Figure 5.4. In our framework, the dataset is built, taking the possible emitter position $(X_e, Y_e, Z_e)$ as features and the peaks (amplitude and delay) in the VCC as the features to be used by the Random Forest classification.

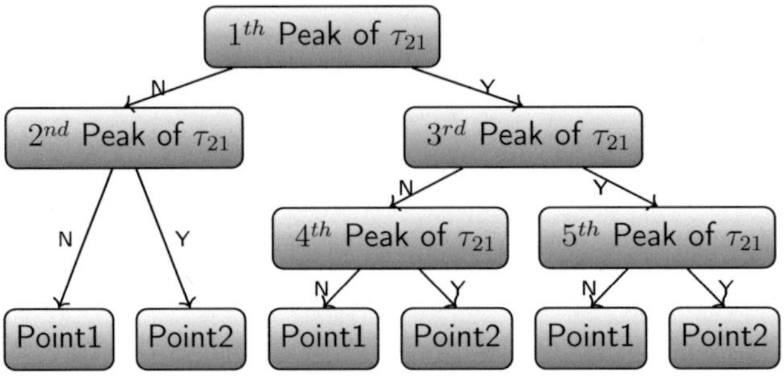


Figure 5.4.: Example of one decision tree for position estimation using RF.

Random Forest in a classification framework produces several decision trees randomly distributed using the features in the dataset. At the end of the classification process is one position defined as a label in the dataset formulation.

The emitter position estimation is done with every "set" of data from the VCC "peaks." The algorithm makes an approximation to find a position as a classification process $(f)$.

This function should be able to map the input variables $(\Delta_{21}, \Delta_{31}, \Delta_{41})$, to an output $(X_e, Y_e, Z_e)$, that is the emitter position.

$$[X_e, Y_e, Z_e] = f(\Delta_{21}, \Delta_{31}, \Delta_{41}). \tag{5.1}$$

The idea is to make a fingerprint that improves continuously with an anchor emitter that performs a recursive averaging and recalculation of the multipath fingerprint patterns.

The ray description should be good enough to establish the model but can not be as precise that it loses the generalization features. This effect is well-known as *over-fitting* and *under-fitting* of the dataset [Zafari et al., 2019].

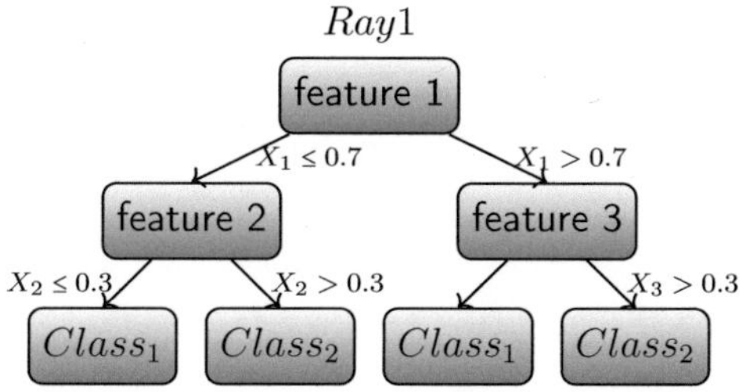


Figure 5.5.: Partition search in Decision Tree for position estimation.

As an example, Figure 5.5 illustrates a decision tree partitioning the input space $\boldsymbol{X} = \boldsymbol{X}_1 \times \boldsymbol{X}_2 = [0;1] \times [0;1]$ for a binary classification problem. The emitter can be in the Region 1, Region 2 or Region 3. The classification subspace is $\boldsymbol{Y} = \{C_1, C_2\}$. The subareas are mapped following the tree structure build with the features of the datasets. The input space $\boldsymbol{X}_1 = \boldsymbol{X}$ is labeled with the binary split $X_1 \leq 0.7$.

Finally, let us observe a binary decision tree applied in a localization problem illustrated in Figure 5.6. The blue dots are related to class $C_1$, and the red dots are related to class $C_2$. We can map the classes to sub-regions; as we can see, the binary tree can assign the classes to wrong regions causing false negative and false positive as described Appendix B.

## 5.5. Formulation of Random Forests

In this section, we present the *Random Forest* formulation, where we first describe the bias-variance decomposition of the prediction error, showing the effect of aggregating randomized models through prediction error reduction and its impacts in decreasing the variance. We also outline the main properties and features of Random Forests.

Machine learning Fingerprint strategies can use single classifiers, such as K-Nearest-Neighbors (KNN), Dense Neural Networks (DNN), and Support Vector Machine (SVM) or decision trees classifiers like Random Forest.

Based on the investigations presented by [Wang et al., 2018] and [Guo et al., 2018], we have proposed a framework using the Random Forest algorithm in both classification and regression techniques to build a cooperative supervised localization model.

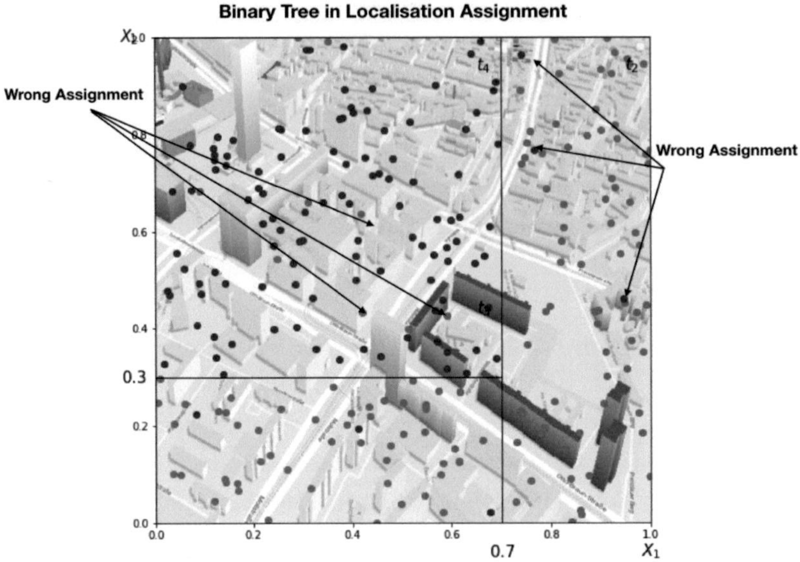


Figure 5.6.: Decision Tree for Localization. Blue dots correspond to objects of class $C_1$ while red dots are objects of class $C_2$ (Figure inspired in [Louppe, 2014] studies).

Random Forest (RF) is an ensemble classifier built using many decision trees (DTs) made with a random selection of the dataset features. In localization problems, it presents fast matching speed and high classification accuracy. The name "Random Forests" stands for several trees grown randomly using the features of the datasets.

Following the statistical formulation carried out by [Louppe, 2014], the input random variables $X_1, ..., X_p$ are mapped in the output variable $Y$. The values of input-output pairs, $X \times Y$, can be defined using the joint probability distribution $P(\boldsymbol{X}, \boldsymbol{Y})$, where $\boldsymbol{X}$ denotes the random vector $(X_1, ..., X_p)$.

The probability that random variables $X$ and $Y$ take values $\mathbf{x}$ and $\mathbf{y}$ from $X$ and $Y$ is $P(X = \mathbf{x}, Y = \mathbf{y})$.

Let us assume an algorithm $A$ trained by the input-output relation $\varphi_L$. The learning model $\varphi$ is the model that minimizes its expected prediction error.

Assuming that $E_X\{f(X)\}$ is the expected value of $f(x)$ with respect to the probability distribution of the random variable $X$. This expected value can also be defined as:

$$E_X\{f(X)\} = \sum_{x \in X} P(X = x) f(x). \tag{5.2}$$

The *expected prediction error* of the model $\varphi_L$ can be formulated as:

$$Err(\varphi_L) = E_{X,Y}\{L(Y, \varphi_L(X))\}, \tag{5.3}$$

where $L$ is the learning set used to build $\varphi_L$ and $L$ is a loss function measuring the discrepancy between its two arguments.

Equation (5.3) is also known as prediction error of $\varphi_L$ over all possible input-output combination, represented by a couple $(\mathbf{x}, \mathbf{y}) \in X \times Y$.

The set of all input-output possible combination is referred to as the sample space, which includes not only the observed couples from the learning datasets $L$ but also all the *unseen* ones from $X \times Y$.

The machine learning formulation does not try to make the most accurate predictions over the subset $L$ of known data, but rather to build a model from the learning process that can predict the output properly using all possible data.

Taking the model in $X$, the output as $Y$, and the machine learning model as $\varphi_L$, we have the prediction equation as follows:

$$Y = \varphi_L(X). \tag{5.4}$$

The general formulation for the expected prediction error associated with a loss function $L$ is given by $Err(\varphi_L)$ in Equation (5.3). The expected error of the predictions of $\varphi_L$ around value $X = \mathbf{x}$ is:

$$Err(\varphi_L(\mathbf{x})) = E_{Y|X=\mathbf{x}}\{L(Y, \varphi_L(\mathbf{x}))\}. \tag{5.5}$$

In the case of the regression approach, the formulation Equation (5.5) can be used for the squared error loss and decomposed into bias and variance terms which together constitute a valuable framework for diagnosing the prediction error of a model.

A similar equation for classification formulation is challenging to obtain. [Louppe, 2014] recommends using bias and variance concepts for comparing the prediction error of different classifiers. Assuming that the multipath pattern can be extracted using RT simulation as presented in Section 4.5.2 and Section 4.5.2, each instance of the dataset can be represented by $N$ Features as described in Figure 5.7. Position Fingerprints using Random Forest can take the multipath data produced by RT to overcome the nonlinearities of outdoor NLOS environments.

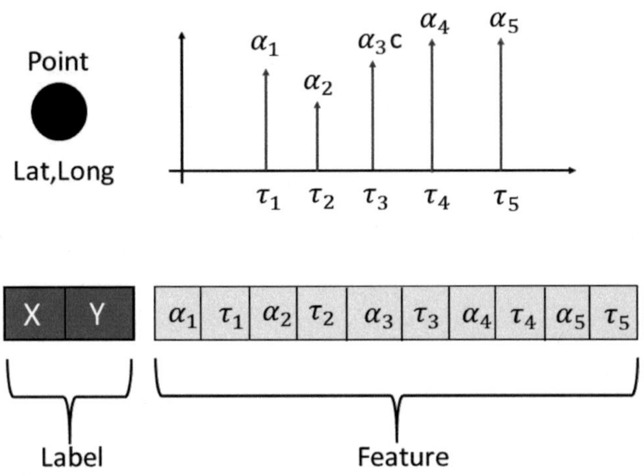


Figure 5.7.: One Instance in the Multipath Fingerprint Dataset.

### 5.5.1. Regression using Random Forest

Assuming that there are $M$ decision trees in the Random Forest model, randomized and built using the dataset features. The expected error for a set of estimations or ensemble error is $\psi_{,\theta_1,\ldots,\theta_M}$. The model can be decomposed in a sum of three terms: noise($\mathbf{x}$), bias($\mathbf{x}$) and var($\mathbf{x}$).

The most used loss function for regression formulation is the *squared error* loss $L(Y, \varphi_L(X)) = (Y - \varphi_L(X))^2$. In this approach makes the large differences between the true values and the predicted values are more costly than the small ones. The generalization error of the regression models is:

$$Err(\varphi_L) = E_{X,Y}\{(Y - \varphi_L(X))^2\} \tag{5.6}$$

Using the formulation of [Louppe, 2014], the model expected prediction error $\varphi_L$, defined in Equation (5.4), we can include the Bayes formulation for the prediction error as $\varphi_B$.

The difference in the prediction can also be modeled as a "bias" noise term that depends on the intrinsic random factor of $Y$. This value is independent of the learning algorithm and stays the same

$$\text{noise}(\mathbf{x}) = E_{Y|X=\mathbf{x}}\{(Y - \varphi_B(\mathbf{x}))^2\}. \tag{5.7}$$

As described by [Louppe, 2014], the ensemble of the predictions error using Bayes and the average of the prediction error are the same. With $M$ decision trees, each one built with a randomized combination of the features is modeled as:

$$\begin{aligned} E_{L,\theta_1,\ldots,\theta_M}\{\psi_{L,\theta_1,\ldots,\theta_M}(\mathbf{x})\} &= E_{L,\theta_1,\ldots,\theta_M}\{\frac{1}{M}\sum_{m=1}^{M}\varphi_{L,\theta_m}(\mathbf{x})\} \\ &= \frac{1}{M}\sum_{m=1}^{M} E_{L,\theta_m}\{\varphi_{L,\theta_m}(\mathbf{x})\} \\ &= \mu_{L,\theta}(\mathbf{x}) \end{aligned} \tag{5.8}$$

If the trees are randomized properly, variables $\theta_m$ are independent from each other but shares the same distribution. It follows that:

$$\text{bias}^2(\mathbf{x}) = (\varphi_B(\mathbf{x}) - \mu_{L,\theta}(\mathbf{x}))^2. \tag{5.9}$$

Equation (5.9) means that the bias in the ensemble of randomized models is equal to the bias of any of the randomized models, meaning that building randomized decision trees does not affect the bias.

The point now is to analyze the variance of the randomized ensemble. Initially, we define the correlation coefficient $\rho(\mathbf{x})$ between the predictions of two randomized models built on the same learning dataset, $\theta'$ and $\theta''$.

$$\rho(\mathbf{x}) = \frac{E_{L,\theta',\theta''}\{(\varphi_{L,\theta'}(\mathbf{x}) - \mu_{L,\theta'}(\mathbf{x}))(\varphi_{L,\theta''}(\mathbf{x}) - \mu_{L,\theta''}(\mathbf{x}))\}}{\sigma_{L,\theta'}(\mathbf{x})\sigma_{L,\theta''}(\mathbf{x})}$$
$$= \frac{E_{L,\theta',\theta''}\{\varphi_{L,\theta'}(\mathbf{x})\varphi_{L,\theta''}(\mathbf{x})\} - \mu^2_{L,\theta}(\mathbf{x})}{\sigma^2_{L,\theta}(\mathbf{x})} \tag{5.10}$$

If the correlation $\rho(\mathbf{x})$ is $1$; the predictions are highly correlated; on the other hand, when it is close to $0$, predictions of the randomized models are decorrelated.

The ensemble variance can be approximated by:

$$\text{var}(\mathbf{x}) = \rho(\mathbf{x})\sigma^2_{L,\theta}(\mathbf{x}) + \frac{1 - \rho(\mathbf{x})}{M}\sigma^2_{L,\theta}(\mathbf{x}). \tag{5.11}$$

The benefit of variance reduction is directly related to the correlation coefficient and the number of decision trees used. For large ensembles ($M \to \infty$), the variance of the ensemble reduces to $\rho(\mathbf{x})\sigma^2_{L,\theta}(\mathbf{x})$.

The randomization can improve the predictions of randomized models, in case of $\rho(\mathbf{x}) < 1$, the variance of an ensemble is smaller than the variance of an individual model. As derived by [Louppe, 2014], [?], [Breiman, 2004], the noise, bias and variance assume the following structure: For the squared error loss, the bias-variance decomposition of the expected generalization error $E_L\{Err(\psi_{L,\theta_1,\ldots,\theta_M}(\mathbf{x}))\}$ at $X = \mathbf{x}$ of an ensemble of $M$ randomized models $\varphi_{L,\theta_m}$ is

$$E_L\{Err(\psi_{L,\theta_1,\ldots,\theta_M}(\mathbf{x}))\} = \text{noise}(\mathbf{x}) + \text{bias}^2(\mathbf{x}) + \text{var}(\mathbf{x}), \tag{5.12}$$

where

$$\begin{aligned}
\text{noise}(\mathbf{x}) &= Err(\varphi_B(\mathbf{x})),\\
\text{bias}^2(\mathbf{x}) &= (\varphi_B(\mathbf{x}) - E_{L,\theta}\{\varphi_{L,\theta}(\mathbf{x})\})^2,\\
\text{var}(\mathbf{x}) &= \rho(\mathbf{x})\sigma^2_{L,\theta}(\mathbf{x}) + \frac{1 - \rho(\mathbf{x})}{M}\sigma^2_{L,\theta}(\mathbf{x}).
\end{aligned}$$

In short, the Random Forest method in regression configuration can theoretically give an estimation error reduction depending on the size of the ensemble and the uncorrelation between the decision trees generated.

### 5.5.2. Classification using Random Forest

It is possible to use a Machine Learning classification procedure to obtain a prediction model. In case of a TDoA localization system is deployed and the multipath fingerprints dataset is available, the model created can estimate the emitter position based on features fingerprints.

As described by [Louppe, 2014], in classification, the most common loss function is the *zero-one* loss function defined as:

$$1(\textit{condition}) = \begin{cases} 1 & \text{if } \textit{condition} \text{ is true;} \\ 0 & \text{if } \textit{condition} \text{ is false.} \end{cases}$$

$L(Y, \varphi_L(X)) = 1(Y \neq \varphi_L(X))$, where all misclassifications are equally penalized. In this case, the generalization error of $\varphi_L$ becomes the probability of misclassification of the model:

$$Err(\varphi_L) = E_{X,Y}\{1(Y \neq \varphi_L(X))\} = P(Y \neq \varphi_L(X)) \tag{5.13}$$

For the TDoA system with stations nodes $S_{i=1:N}$, each possible emitter position can be a class $C_i$, making a mapping of the CIR features to each point. As a result, we can determine the position of the emitter.

If we define a set of classes $C_1, C_2, \ldots$, with each class $C_i$ being a geographical region in where the emitter might be located. The size of each class depends on the grid level or the discretization assumed in the dataset formulation; throughout our research, we have to keep the grid size compatible with the collecting region of the RT simulation concept introduced in Section 4.2.

The Random Forest model in classification approach uses a direct analogy with the bias-variance decomposition of the squared loss error derived for regression. The difference in classification is the loss function that is a "zero-one" type, meaning either the class was matched or not. Therefore, with $M$ randomized decision trees, the expected error for a set of estimations or ensemble error is:

$$E_L\{E_{Y|X=\mathbf{x}}\{1(\varphi_L(x) \neq Y)\}\} = P_{L,Y|X=\mathbf{x}}(\varphi_L(\mathbf{x}) \neq Y) \tag{5.14}$$

As discussed by [Dietterich and Kong, 1995] and [Dietterich, 2000], in direct analogy with the bias-variance decomposition for the squared error loss, it is necessary to redefine the concepts of bias and variance in the case of classification. The estimates in classification can be computed as follows:

$$\hat{p}_L(Y = c|X = \mathbf{x}) \tag{5.15}$$

The conditional probability of a class derived from a decision tree is: $(\hat{p}_L(Y = c|X = \mathbf{x}) = p(c|t))$. In our fingerprint approach a prediction with classification models is the operation to find the class that maximizes the estimate, as:

$$\varphi_L(\mathbf{x}) = \underset{c \in Y}{\operatorname{argmax}} \{\hat{p}_L(Y = c|X = \mathbf{x})\} \tag{5.16}$$

The derivation of bias and variance using classification models can be done using the numerical estimates of the expected classification error. Therefore, only with a set of classifications, a successful rate can be extracted, given the bias and the variance of the estimations.

Here, it is fundamental to approach the problem statistically, with the confusion matrix presented in Table B.1, associating the random nature of the output $Y$. The expected value wrong classification can be represented by:

$$E_L\{E_{Y|X=\mathbf{x}}\{1(\varphi_L(\mathbf{x}) \neq Y)\}\} = P_L(\varphi_L(\mathbf{x}) \neq Y), \quad \text{or,} \tag{5.17}$$

$$E_L\{E_{Y|X=\mathbf{x}}\{1(\varphi_L(\mathbf{x}) \neq Y)\}\} = 1 - P_L(\varphi_L(\mathbf{x}) = Y). \tag{5.18}$$

Hereafter, we reproduce the derivation presented by [Louppe, 2014], using the Bayes model to find a decision boundary, as follows:

$$
\begin{aligned}
E_L\{E_{Y|X=\mathbf{x}}\{1(\varphi_L(\mathbf{x}) \neq Y)\}\} &= 1 - P_L(\varphi_L(\mathbf{x}) = \varphi_B(\mathbf{x}))P(\varphi_B(\mathbf{x}) = Y) \\
&\quad - P_L(\varphi_L(\mathbf{x}) \neq \varphi_B(\mathbf{x}))P(\varphi_B(\mathbf{x}) \neq Y) \\
&= P(\varphi_B(\mathbf{x}) \neq Y) + P_L(\varphi_L(\mathbf{x}) \neq \varphi_B(\mathbf{x})) \\
&\quad - 2P_L(\varphi_L(\mathbf{x}) \neq \varphi_B(\mathbf{x}))P(\varphi_B(\mathbf{x}) \neq Y)
\end{aligned}
$$

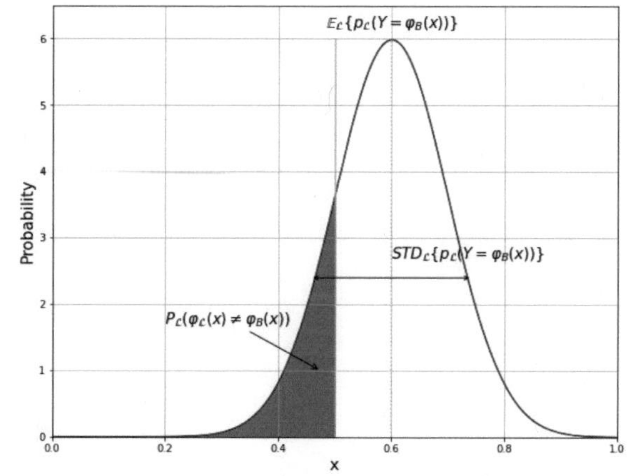


Figure 5.8.: Probability distribution of the estimate $\hat{p}_L(Y = \varphi_B(\mathbf{x}))$, modified from [Louppe, 2014].

$$
\begin{aligned}
E_L\{E_{Y|X=\mathbf{x}}\{1(\varphi_L(\mathbf{x}) \neq Y)\}\} &= P(\varphi_B(\mathbf{x}) \neq Y) \\
&\quad + P_L(\varphi_L(\mathbf{x}) \neq \varphi_B(\mathbf{x}))(2P(\varphi_B(\mathbf{x}) = Y) - 1)
\end{aligned}
\tag{5.19}
$$

The first expression in Equation (5.19) represents the Bayes model, and the second term represents the increased error due to the wrong estimation, assuming the optimal decision boundary. The probability of the model decision, differently from the Bayes model, is $P_L(\varphi_L(\mathbf{x}) \neq \varphi_B(\mathbf{x}))$.

As represented schematically in Figure 5.8, the probability (5.20) corresponds to the tail area on the left side of the decision threshold (at 0.5) of the distribution. The estimation is different from the Bayes model when $\hat{p}_L(Y = \varphi_B(\mathbf{x}))$ is lower than $0.5$, which is represented by:

$$P_L(\varphi_L(\mathbf{x}) \neq \varphi_B(\mathbf{x})) = P_L(\hat{p}_L(Y = \varphi_B(\mathbf{x})) < 0.5) \tag{5.20}$$

For a normal distribution of the estimates $\hat{p}_L(Y = \varphi_B(\mathbf{x}))$, the probability can be calculated from the mean and variance using Equation (5.20), as follows.

$$P_L(\hat{p}_L(Y = \varphi_B(\mathbf{x})) < 0.5) = \Phi\left(\frac{0.5 - E_L\{\hat{p}_L(Y = \varphi_B(\mathbf{x}))\}}{\sqrt{V_L\{\hat{p}_L(Y = \varphi_B(\mathbf{x}))\}}}\right) \tag{5.21}$$

The cumulative distribution function of the standard normal distribution [Walck et al., 2007], is given as:

$$\Phi(x) = \frac{1}{\sqrt{2\pi}} \int_{-\infty}^{x} \exp\left(-\frac{t^2}{2}\right) dt. \tag{5.22}$$

In summary, the Random Forest of several decision trees, randomly chosen based on the features, will have an overall effect in reducing the variance of the expected generalization error. When the expected probability estimate $E_L\{\hat{p}_L(Y = \varphi_B(\mathbf{x})\}$ for the true majority class is greater than $0.5$, a reduction of the variance of the estimate results in a decrease of the total misclassification error. If $V_L\{p_L(Y = \varphi_B(\mathbf{x}))\} \to 0$, then $\Phi \to 0$ and the expected generalization error tends to the error of the Bayes model. The Random Forest method, in regression configuration, can theoretically give an estimation error reduction that depends on the size of the ensemble and the uncorrelation between the decision trees generated.

## 5.6. Evaluation of Multipath Fingerprints using Random Forest Models

This section presents a preliminary evaluation to demonstrate the performance of the localization approach using machine learning and RT multipath fingerprints in the radio-frequency emitter localization problem.

The main overhead for this approach is the training procedure that was carried out using only RT simulations. Algorithm 2 shows the general steps to perform the RT

Multipath Fingerprints estimation with the Random Forest in classification approach.

---

**Algorithm 2:** RF Fingerprint in TDOA Systems.

---

**Input** : Similar CBB signal received by TDoA System
RT Simulations (TDoA Sensor Position Buiding Description)
**Output:** Estimated Emitter Position
**for** $P_i \in \Delta_x : \Delta_y$ *in RT Grid* **do**
From a deployed TDOA Sytem:;
· Record the signal, extract the CBB;
· Extract VCC TDOA with 5 Peaks $[\tau_{21}, \tau_{31}, \tau_{41}]$;
With RT output collect Produce a VCC Fingerprints;
· 5 $(\alpha_i, \Delta_{\tau_i})$ in each TDOA;
Build a Classification Random Forest:;
· 30 Features: 5 Peaks $(\alpha_i, \Delta_\tau i)$ for VCC TDOAs;
· 3 Outputs: Emitter Position $[\mathrm{X_e}, \mathrm{Y_e}, \mathrm{Z_e}]^\mathrm{T}$;
Refine the Dataset of RF:;
· Record from the Calibration;
· Perform VCC CIR extract reference;
VCC TDOAS from calibration points;
· Perform "reinforcement";
· Adjust the dataset of Random Forest Engine;
Perform the Position Estimation;
Apply Random Forest method and refined the dataset;
Repeat the algorithm with a new integration windows until the end of the recording signal.;
Perform the joint position estimation with: $P(\mu_{TDOA}, \sigma_{TDOA}) \cap P(\mu_{RT}, \sigma_{RT})$
**end**

---

The evaluation of Multipath Fingerprints using Random Forest Models was performed at the TU Ilmenau campus, Germany, with CIR and VCC datasets. The same scenario was used to compare regression and classification approaches. The area was discretized in a grid of 2 m for the possible emitter positions, and the TDoA sensor positions were defined in the RT simulation, generating the multipath dataset fingerprints.

In Figure 5.9, we show the location error results for the Random Forest Regression approach. We have applied a modified framework, presented by [Yun and Iskander, 2015]. The number of decision trees was set for twelve (12) using a randomized choice of the features set.

The blue dots are the actual emitter positions, while the green dots are the estimations; the red lines connect the estimated points to the actual emitter position. The values follow the RT fingerprints strategy proposed in Section 4.5.1 and Section 4.5.2.

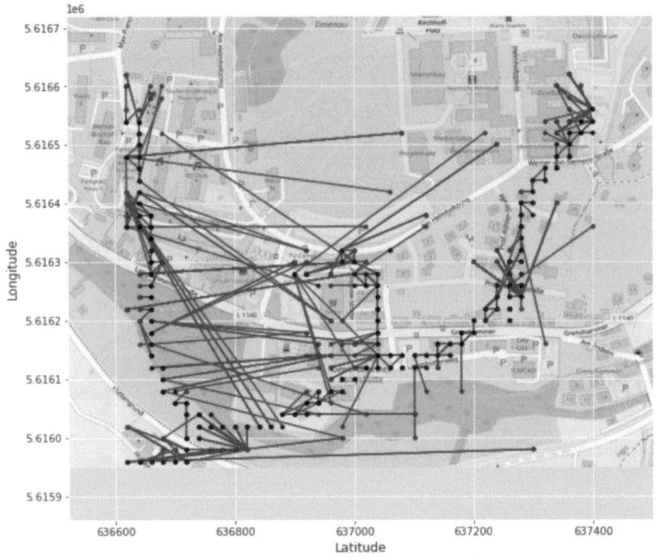


Figure 5.9.: Random Forest Prediction using Regression.

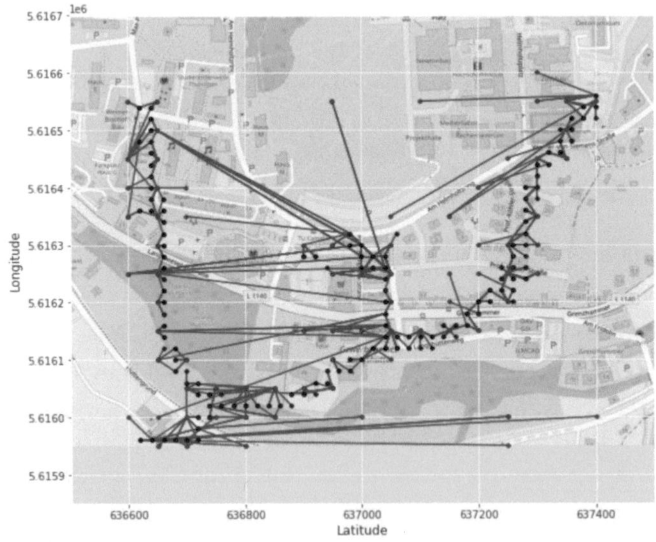


Figure 5.10.: Random Forest Prediction using Classification.

Figure 5.10 shows the prediction performance using Random Forest in a classification approach. We can observe that the errors are lower than the regression implementation.

Some points present a substantial deviation from the actual emitter positions in both regression and classification. The multipath components affected not only the fingerprints approach but also the real measurements.

Figure 5.11 shows the simulation using the same RT multipath fingerprints datasets. This time real measurements were added to the dataset allowing the **data augmentation** procedure as explained by [Yu et al., 2017] and [Sinha and Hwang, 2020]. The model used was the Random Forest in Classification approach.

Keep in mind that in our case, the data augmentation is to assign more information (features) to the same class (label) in the classification procedure. The overall effect is to improve the random decision trees turning the variance of the model smaller.

As a result, the blue dots are the actual points, the green ones are the predicted ones, and the length of the red line represents the Euclidean error. There are points where the Root Mean Squared Error (RMSE) error is high. However, it is still valid once a fusion localization system can give a rough estimation. It integrates it with some other sensors fusion [Liggins II et al., 2017] approach to enhance the overall localization performance.

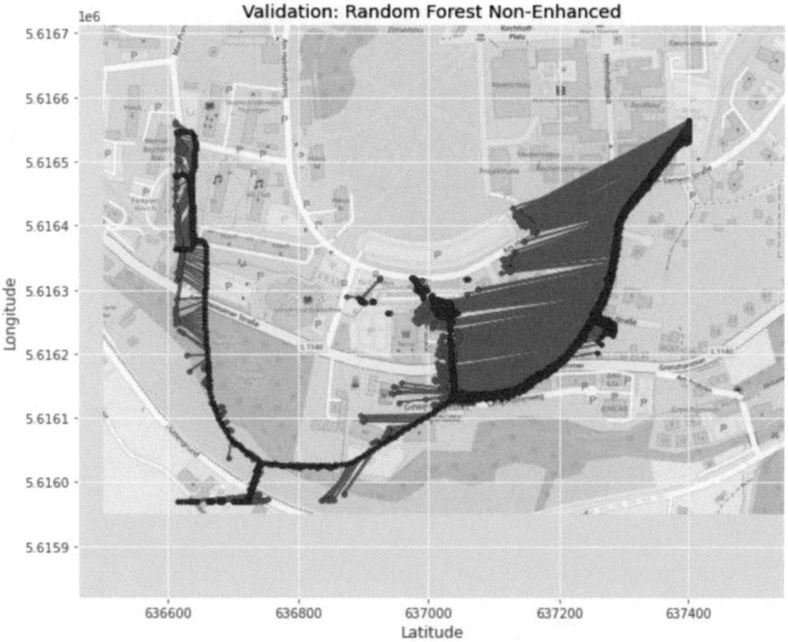


Figure 5.11.: Data Augmentation Random Forest Prediction.

## 5.7. Noise and Mismatching Effects in the Prediction

This section analyses the effects of noise and the mismatching between the measured values, taken from TDoA sensors, and the synthetic one based on RT simulation.

Because of practical limitations, real-world data is often imperfect or incomplete. The dataset produced with RT simulations is also imperfect because there are several approximations on the scenario description and the material characteristics included.

Figure 5.12 shows the effects of the mismatching and noise schematically. Noise causes incorrect values to be taken as features; more explicitly, the amplitude or the delay are affected by real-world imperfections leading to values that differ from reality. Mismatching is a different effect caused by the wrong ray assignment, which happens when we have less features than the model, or the features are assigned in the wrong way. In this case, a multipath component in the model and does not exist in the measurements causes the mismatching error.

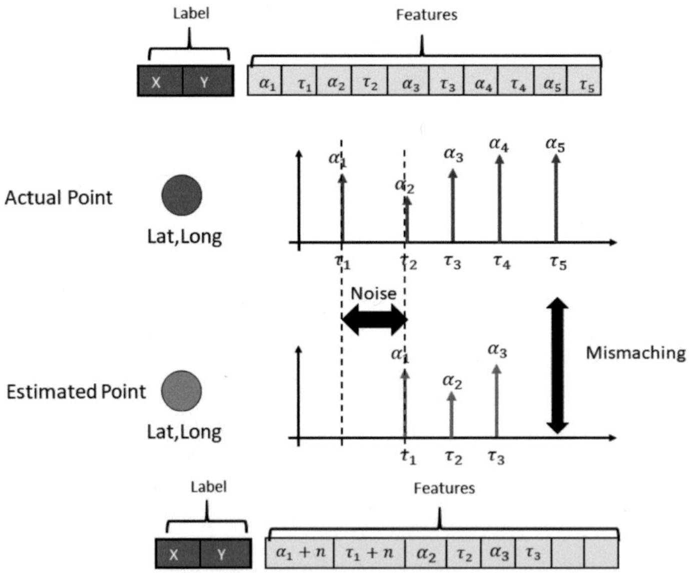


Figure 5.12.: Noise and Mismatching in Random Forest.

The studies carried out by [Breiman, 1996] and [Breiman, 2004] showed that the Random Forest model has several strategies to deal with mismatching, noise, and missing data in the dataset.

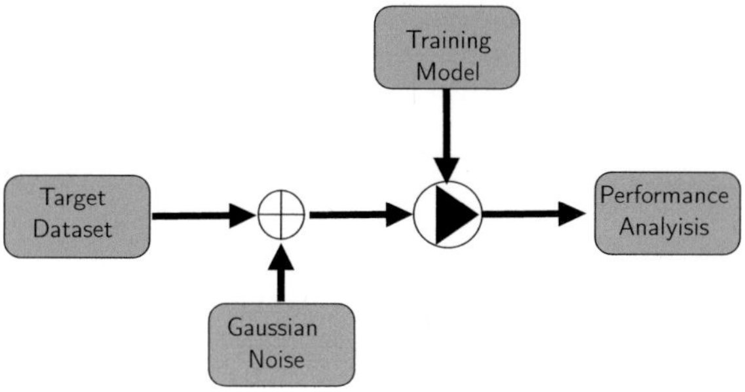


Figure 5.13.: Noise Experiment in Random Forest

The most straightforward approach is the so-called ternary decision trees, which is an approach to include the missing values on the decision trees. The trees are transformed to explicitly model the missing values as ternary splits instead of binary splits.

The second approach is to propagate the missing values to both sides of the decision trees, then the value of the split variable is missing into both left and right child nodes.

The last approach is called imputation, where the Random Forest algorithm includes some values in the missing features. It can be done using several mechanisms for imputing missing values. The straightforward approach uses a rough and inaccurate approximation, taking the median of the neighbor features.

We have carried out some simulations to observe the robustness of the Random Forest against noise and mismatching on the measurements.

We have added zero-mean Gaussian noise in all features and a defined variance of each feature of all instances from the dataset. The $level$ was from 0.1 to 1.0, with an increment of 0.05, and from 1 to 9, with an increment of 1. Figure 5.13 shows the Noise Experiment Process.

Figure 5.14 shows the results of this experiment. The black line in the middle is the mean Euclidean distance error between real and predicted positions of all blue dot points using Random Forest Classification. Assuming that $\mu$, as mean and $\sigma$, as the standard deviation of the errors in the positioning, the orange area defines $\mu - \sigma$ and $\mu + \sigma$ limits.

We may notice that the standard deviation from Euclidean distance error is high, which means several positions suffer from noise. The principal cause of the high values is

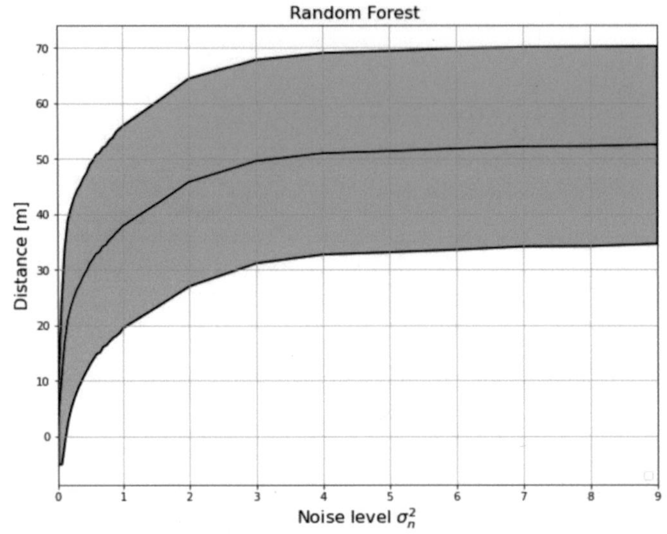


Figure 5.14.: Performance of Random Forest Prediction with noise.

that the model was generated using simulation data and executed on real data (another data distribution). However, several estimated locations were obtained with an error of lower than 50 meters.

Our dataset has five peak values of the VCC or five Specular Components from each TDoA sensor. A subset of features is nullified in the mismatching experiment, and the model's performance is obtained. The selected noise level for this experiment was fixed. Figure 5.15 presents this process.

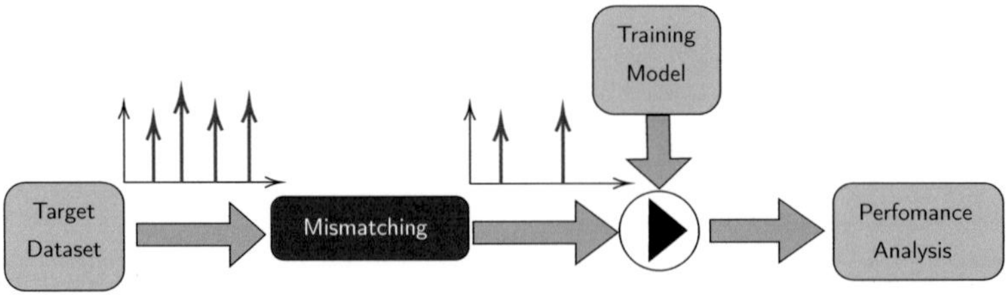


Figure 5.15.: Mismatching Experiment in Random Forest.

Since the TDoA system with $4$ sensors, the features are parameter pairs of the amplitude and the delay - $\alpha_i, \tau_i$. We have selected the $n$ most significant pair of features, where $n$ is in the range from 1 to 20.

Figure 5.15 shows the setup to analyze the ability of the predictor to deal with missing data in the dataset. Figure 5.16 presents the performance for the Random Forest model when the number of feature varieties from $1$ (one amplitude and one delay) to $20$. Again, the middle black line is the mean Euclidean distance error between real and predicted positions using Random Forest Classification. The red area defines $\mu - \sigma$ and $\mu + \sigma$ limit.

There is a steady value where there is no improvement with more features. The reason is that many features in the simulation dataset are already zero.

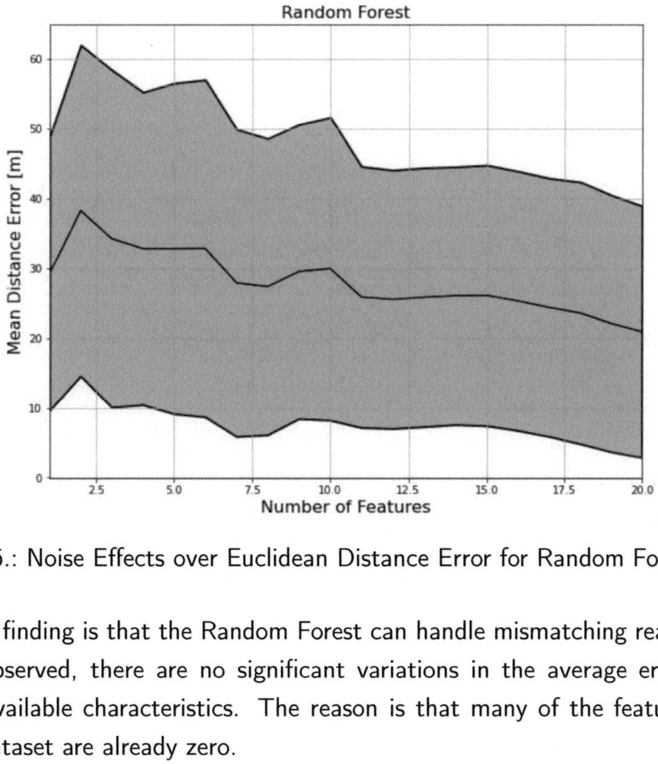


Figure 5.16.: Noise Effects over Euclidean Distance Error for Random Forest model.

The first finding is that the Random Forest can handle mismatching reasonably well; as can be observed, there are no significant variations in the average errors with the number of available characteristics. The reason is that many of the features from the simulation dataset are already zero.

## 5.8. Random Forest Implementation

This section explains general aspects of the Random Forest models applied to build the RT Multipath Fingerprints structure proposed in this research. A Random Forest is an estimator that fits several decision tree classifiers on various sub-samples of the dataset and uses averaging to improve the predictive accuracy. Several aspects need to be considered to set the Random Forest parameters. Therefore, it is appropriate to explain the most relevant ones to the framework proposed.

Our simulation scripts used the Scikit-Learn machine learning $Python$ library, which offers a collection of machine learning tools for Python. It includes algorithms, pre-processing tools, and model selection procedures. Scikit-Learn combines numeric and scientific packages based on NumPy and SciPy libraries, offering a unified mechanism to produce the machine learning workflow.

Most of the Random Forest parameters are default, but we should have special care in the definition of the *number of trees* or *number of estimators*. This parameter represents the random trees used in the classification forest. In general, by increasing the number of trees, the results improve as well. However, there is a saturation point where we can not improve, increasing the number of trees. At this point, the benefit in prediction performance from learning more trees will be lower than the cost in computation time for learning these new trees.

Random Forests are ensemble methods. It means that they average the prediction across many trees. The expected variance tends to decrease as the square root of the sample size. At a certain point, the cost of collecting a larger sample size will be higher than the benefit in accuracy obtained from such a larger sample.

There are some guidelines to determine the number of trees [Biau and Scornet, 2016]. Using the square root of the total number of input features is recommended as a rule of thumb criteria.

The Figure 5.17 shows the prediction performance using changing the number of estimators. We can observe that after $13$ estimators, the Euclidean error stays around $37$ meters.

It is possible to use parallel computing using the internal Scikit-Learn Random Forests features, which supports distributed processing on multi-core architectures. The simple approach is to assign each estimator, representing each tree to one different processor, summing up the iterative results using Message Passing Interface (MPI), [Mitchell et al., 2011].

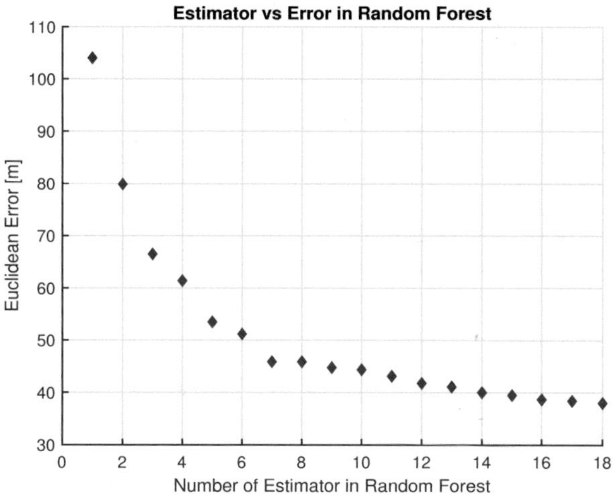


Figure 5.17.: Number of estimator in Random Forest vs Euclidean Error.

The construction of a forest can be a parallel problem, in which the $M$ decision trees are executed independently on multiple processors. If there is no parallelization overhead, the computational speed is equal to the number $N_p$ of processors used for building the forest. Parallel processing is out of the scope of this research, but it is an appealing feature of Random Forest classification to be further investigated.

## 5.9. Concluding Remarks

This chapter illustrates and reviews the use of machine learning in localization problems; furthermore, we have shown how to use it with RT positioning fingerprints. We have investigated the ways to explore multipath information from the CIR in each sensor, as explained in Chapter 3. Additionally, we have presented Machine Learning and Random Forest terminology to make a shared understanding of our general framework.

We have linked the machine learning approach with our RT multipath fingerprints datasets, where we emphasized feature extraction in the outdoor scenario. The Decision Trees implementation and mathematical foundation were explained both in regression and classification approaches. Furthermore, we have demonstrated that we have focused on the Random Forest models due to their capability of processing heterogeneous and high non-linear data.

The Random Forest classifier can be set to operate for both classification and regression configuration. It can handle the missing values that are advantageous in our multipath fingerprint approach because sometimes differences between simulated datasets and the actual values can occur.

The concept of multipath fingerprints was linked to the formulation of Random Forests, where the simulations were performed to evaluate the robustness of the model to noise and mismatching effects. We also discussed critical aspects of the Random Forest implementation, including the theoretical foundations. We have concentrated our studies on the classification models using Random Forest to integrate into our fingerprints approach.

This chapter concludes the first part of the research, which was focused on describing the fundaments of the radio-frequency emitter position with RT multipath exploitation. Localization problems of TDoA system in an outdoor scenario, channel estimation of localization systems, the RT fingerprints to build the datasets, and the machine learning using Random Forest classification are building blocks to understand the framework proposed.

# Part II

# Multipath Position Fingerprints

# 6. Contributions for Multipath Exploitation in RF Outdoor Localization

This chapter explains the main ideas of the research; in it, we analyze the critical aspects presented in the previous chapters and put them together to explain the proposed method. The final goal is to enhance a localization system deployed in an outdoor scenario, facing multipath effects.

Therefore, before starting, it is imperative to establish a clear understanding of the word enhancement. According to the Oxford Dictionary, the term"**exploitation**" means to take as much benefit as possible from some available resource [Dictionary, 1989]. In our case, we try to extract as much information as possible from the multipath effects in a localization scenario.

As described in detail in Section 4.1, several studies attempted to employ this concept; some of them worked solely on the enhancement of multipath description while others integrated it into a hybrid approach for localization. However, in this research, we plan to restrict ourselves to increase the system performance. Consequently, working out a method to integrate machine learning fingerprint estimations with actual measurements to improve the overall accuracy is the ultimate purpose.

This chapter begins by highlighting the contributions to the research mentioned in Section 1.5. We have adopted the Inverse Covariance Intersection (ICI) to integrate machine learning estimations with the actual measurements.Appendix C shows the mathematical formulation, explaining the motivations to adopt this approach in our localization problem. In the second section, the localization enhancement approaches are explained step-by-step. Initially, we are only ray-tracing simulations to extract the first Non-Line Of Sight (NLOS) components in the outdoor scenario. After that, we describe the integration of Volume Cross-Correlation (VCC) fingerprints in the Random Forest model in classification and regression schemes. In the end, we describe the formulation of a combined dataset using Ray Tracing (RT) simulations and Time-Difference Of Arrival (TDoA) systems to enhance position estimation.

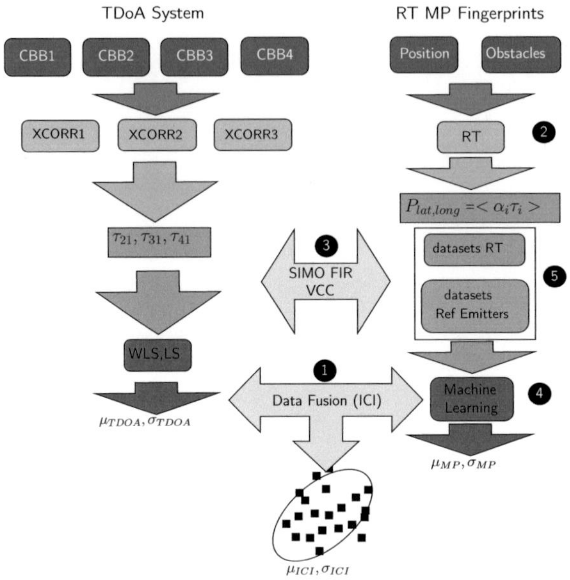


Figure 6.1.: Main Contributions in Multipath Fingerprints Enhancement.

## 6.1. Enhancement Approaches for Localization

This section presents the possible setups to use RT Multipath fingerprints with machine learning. Figure 6.1 on the left side, presents the usual procedure adopted by a TDoA localization systems, as described in detail in Chapter 2. The Complex Baseband (CBB) signal recorded by a set of sensors produces the time-differences, and a Weighted Least Square (WLS) algorithm produces the position estimations. These estimations are affected by multipath, so it is possible to extract the mean and the variance of the set of estimations ($\mu_{TDoA}, \sigma_{TDoA}$).

The right side of Figure 6.1 shows the matching of contributions with the general multipath fingerprint framework, along with the aspects discussed in Section 1.5. The TDoA sensor positions and the obstacle description feed the RT simulation, producing the multipath fingerprints patterns that assign to each point in the scenario a distinct Channel Impulse Response (CIR). The TDoA measurements, either using the VCC or the Single Input Multiple Output (SIMO) FIR of the CBB are compared to the RT fingerprints.

The multipath dataset using RT is enhanced with the values from real measurements. The machine learning engine uses multipath fingerprint patterns to estimates the emitter position. In the end, both estimations are fused using the ICI framework, creating a more precise position prediction $(\mu_{ICI}, \sigma_{ICI})$.

Outdoor localization systems use a statistical approach to present the results. Initially, the real TDoA system shows substantial variations and incorrect estimations scattered open a large area. However, as soon as more measures are drawn, the displayed data is updated. The results of the localization system are given in terms of mean and variance $(\mu, \sigma)$.Therefore, the performance of an existing localization system is a continually updated process providing the *most probable* position for the emitter.

Figure 6.2 illustrates the usual performance of a TDoA localization system. The *magenta dots* represent the predicted localization area for an emitter that is located at the *yellow dot* position. For an accurate estimation it is necessary to apply mean, variance or other such tools of position performance, namely Geometric Dilution of Precision (GDOP) or Circular Error Probability (CEP), (as described in Section A.2 and Section A.3).

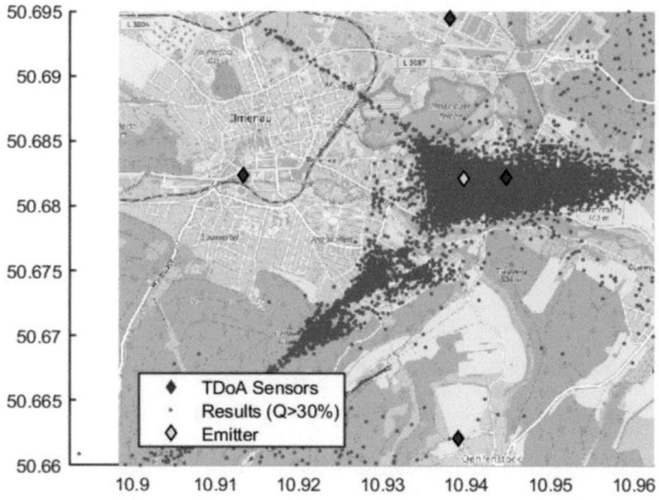


Figure 6.2.: Performance of a Traditional TDoA Localization System in an Outdoor Scenario.

The practical limitations of existing TDoA systems were explained in Section 2.4. It is expected that the approach introduced here can bring some degree of improvement in the overall system performance.

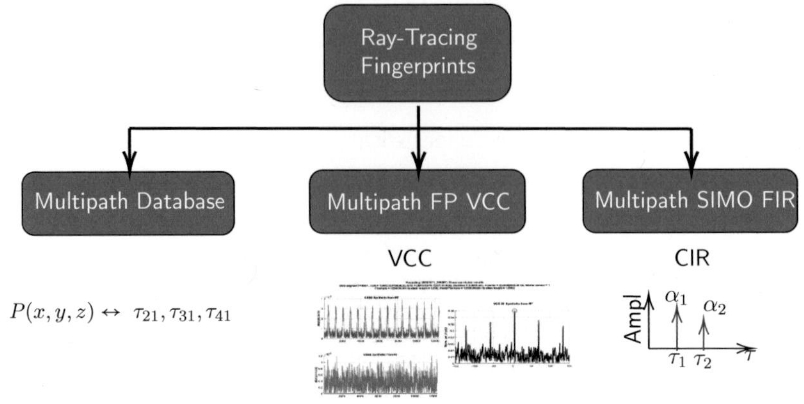


Figure 6.3.: Approaches for Multipath Fingerprints Enhancement.

In the following subsections, we describe the three approaches to exploiting the multipath information in localization problems. Figure 6.3 shows the three main approaches used in our investigations to use multipath information. The multipath database produced only with RT, the multipath fingerprints using VCC and machine learning and using the blind CIR estimation with the SIMO FIR.

### 6.1.1. RT Multipath Database Exploitation

The first proposed strategy to enhance localization performance is to use the quality in TDoA measurements as presented in Section 2.4.4. This parameter is a figure of merit to evaluate if a cross-correlation is "stable" or affected by NLOS situation. Usual TDoA systems filter out the low-quality measurements to improve the overall accuracy.

We can use the mean and the variance of time-differences with RT simulation to produce the TDoA map. The RT simulations can be used to build a TDoA fingerprint of the outdoor scenario; it is possible to assign time differences to every point of the grid that represents the scenario.

The quality factor works as an outlier filter for the TDoA estimation. Usually, the measurements affected by multipath have low-quality values. Hence, if we use only high-quality TDoAs, we filter out the multipath contribution. Figure 6.5 shows the quality

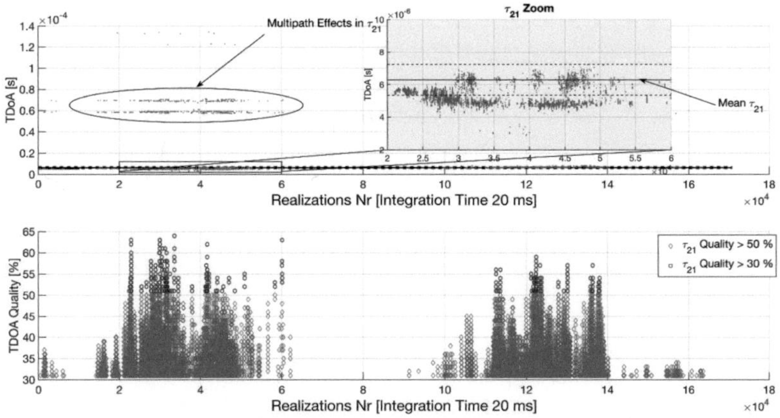


Figure 6.4.: Multipath Effects in Outdoor TDoA.

factor for $\tau_{21}$ produced by a TDoA system. We can observe the enhancement caused by filtering out the low-quality time differences.

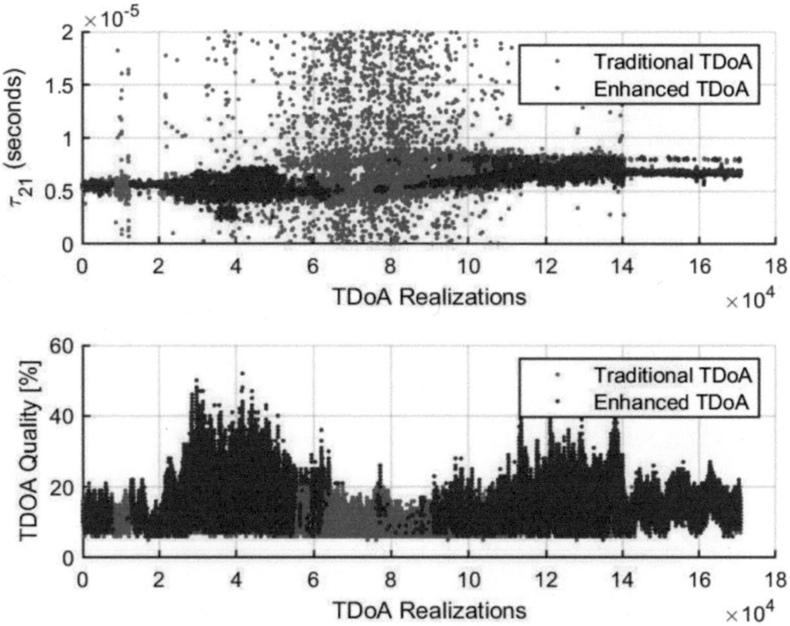


Figure 6.5.: Enhancement with the ICI Fusion of the RT fingerprints and TDoA.

As described by [de Sousa and Thomä, 2018e], it is possible to use sensor position and obstacle description of the region to build a TDoA database.

As indicated by Figure 6.4 multipath disturbs TDoA measurements. It is visible in the zoom that $\tau_{21}$ is characterized by oscillations between $\mu - \sigma$ and $\mu + \sigma$. Our first approach was to take these variations and search the database for the possible position that matches this range of values.Figure 6.6 shows the RT software WinProp [Hoppe et al., 2017] used to generate the dataset as a function of coordinates (Lat,Long), and full 3-D multipath ray data for each point.

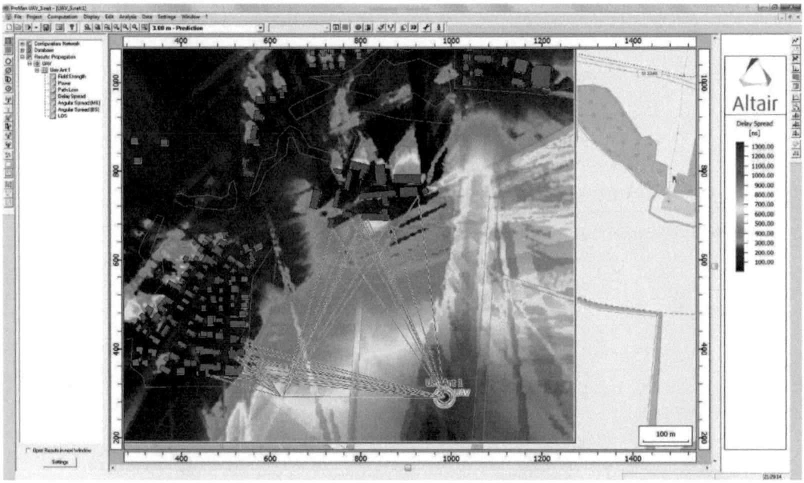


Figure 6.6.: Ray Tracing Simulation to extract the MP Fingerprints.

These multipath patterns represent the RT simulation of the area, including the CIR scenario fingerprints. We can realize in the figure that the NLOS effects producing values that are biased in the measurements.

Figure 6.7 shows the final result of the RT multipath, and each point was assigned in the grid with the mean and variance values of the time-differences measurements using a suitable quality factor. The range of values for the possible TDoA defines a collection of points where the emitter might be. The TDoA database search was used to find the most likely locations in the TDoA range. Each set of possible positions related to the range of values for TDoA generates an estimation that is fused with the real measurements using the ICI approach.

Figure 6.8 illustrates the empirical Cumulative Distribution Function (CDF), where

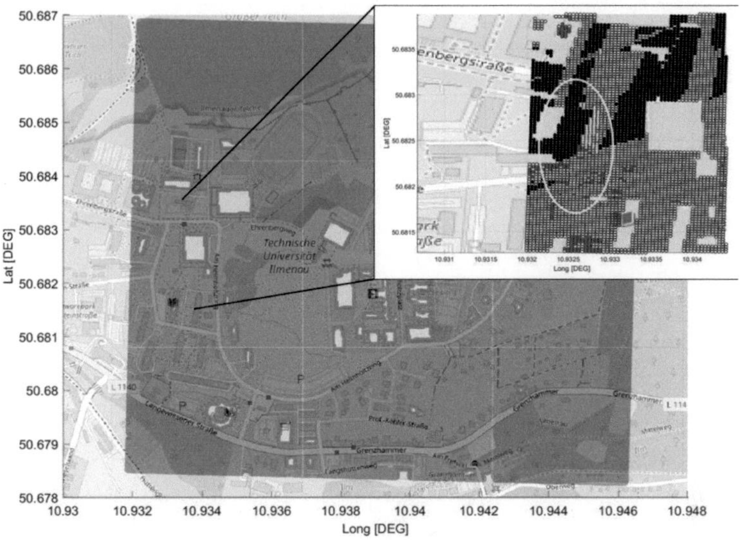


Figure 6.7.: RT Multipath estimation with TDoA Fingerprints intersection.

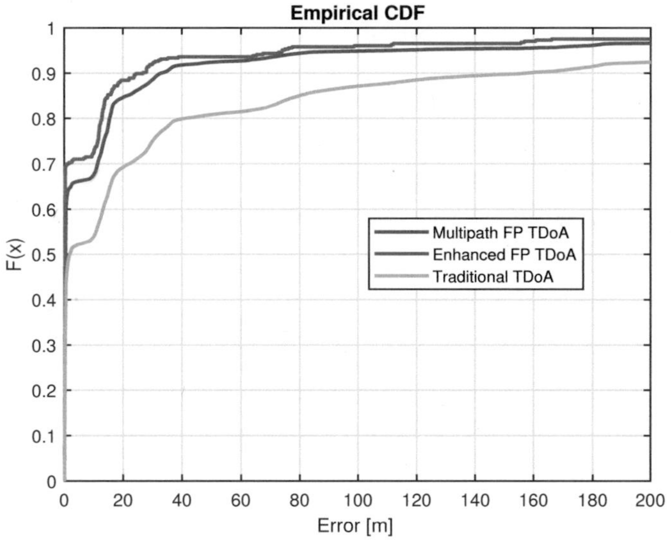


Figure 6.8.: Cumulative Distribution Function of Estimation Error.

once again, it is shown the enhancement achieved by using the proposed method. This approach demonstrated promising results using only RT simulation, thus, serving as a valuable addition to the previously existing information. The improvement can act directly on TDoA results, thereby improving the time difference estimation before computing the position.

### 6.1.2. Matching Real Measurements with RT Multipath

This section deals with the second contribution of this research, the localization multipath fingerprints approach. Figure 6.4 clarifies why it is necessary to match the RT simulation and the actual measurements from a TDoA system.

The signals in each sensor are available in CBB form, along with the cross-correlation among them. As emphasized in Section 2.4.5, this operation is affected by NLOS, producing other peaks that are related to the multipath situation in target-sensor links.

Therefore, we have suggested the application of SIMO FIR for CIR extraction in each CBB, thus, matching measurements with RT simulations. These suggestions were also previously described in [de Sousa et al., 2018], [de Sousa and Thomä, ], [de Sousa and Thomä, 2018b] and [de Sousa et al., 2018].

We have also proposed the direct extraction of multipath fingerprints from the VCC calculated in each CBB pair, as discussed in [de Sousa and Thomä, 2019a] and [de Sousa and Thomä, 2019b]. Furthermore, we have suggested using the other peaks in the cross-correlation function for mapping the multipath pattern at each possible point of the scenario.

The creation of a multipath fingerprint pattern, producing synthetic VCC peaks using RT simulation as a base, has also been suggested. For this, we used the scenario grid CIR and a sample of the received signal to emulate the VCC multipath effects. Figure 6.9 shows the application of the VCC extraction procedure, where we can observe that there are three dominant peaks, but the small peaks can also be seen in the top view figure. Our framework takes only the five strongest peaks information $(\alpha_i, \tau_i)$ to build the multipath fingerprints dataset.

The operation was repeated 251 times, and it is clear that not only the first maximum stays in the same position but also the others peaks, which are related to the multipath composition of the emitter-sensor links, as explained in Section 3.5. The other peaks were either generated synthetically or extracted from the CBB. These findings serve as

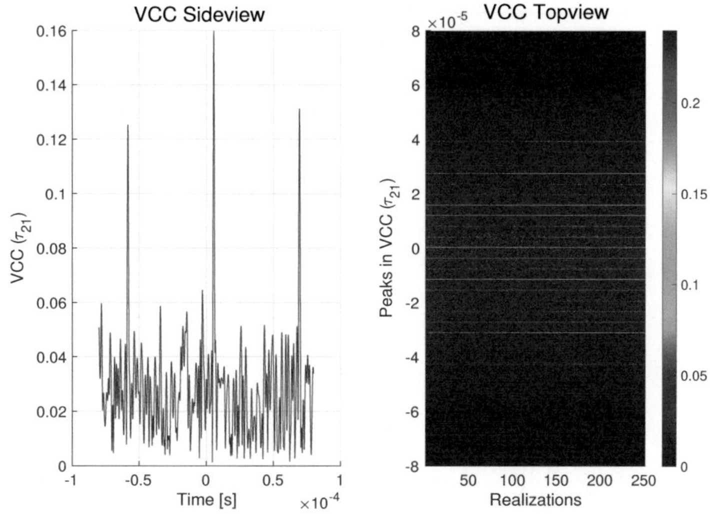


Figure 6.9.: Peaks in VCC representing multipath patterns.

an integration key between the real measurements and the RT simulations.

### 6.1.3. Multipath Fingerprints using Regression NN

This section explains the third main contribution to the outdoor localization problems. We have improved the initial idea by taking into account five peaks in the VCC or five Specular Component (SC) per sensor.

According to this modified approach, five components in each sensor are used to build the features in the machine learning framework, presented in Section 5.6. The fingerprint localization uses RT and the SIMO FIR or VCC to connect the measurement and the synthetic data.

Assuming that, in this scenario, besides TDoA sensors, we also have reference points ($N_r$) and anchor points $N_t$, to aid the estimation of emitter position, which is labelled as an "unknown point" $N_u$. The known positions from anchor nodes are represented by $T_i$ with $i \in 1, \cdots, N_t$; and from the reference nodes by $R_j$ with $j \in 1 \cdots, N_r$. Finally, the emitter positions are unknown nodes, which will be defined by $U_k$ with $k \in 1, \cdots, N_u$.

In our proposal, the anchor nodes transmit their position, and the CIR estimated, consequently, can be compared with the multipath Fingerprint database. The rays are reflected and diffracted by the scenario obstacles, so the propagation environment of interest is assumed to be very complicated. The majority of the first arrivals measured by the TDOA sensors are not due to LOS rays.

A common fingerprinting approach typically uses an off-line calculation, where the multipath fingerprinting is carried out using the reference node signal, where $r_j$ is at the $j^{th}$ reference node position ($j \in 1, \cdots, N_r$).

With this understanding established, the next step is to define a model that can better describe the relationship between the reference node position, the ray-tracing simulations, and CIR estimation.

The model maps the CIR and ray tracing fingerprints as $R_j \approx f(r_j)$ in a vector format. We use as many measurements and reference points as needed to achieve a certain precision level.

The regression models are a class of supervised algorithm [Sohn et al., 2007], that attempts to establish a continuous relationship between a set of dependent variables (geolocation coordinates) and a set of independent variables (multipath fingerprints - $\alpha_i, \tau_i$).

There are some benefits of using ray tracing and real measurements together in the ML engine. Firstly, ML is used to approximate the position as a target function for optimization. This function should be able to map input variables ($X$), which are the multipath components ($\alpha_i, \tau_i$), to output variables ($Y$), which are the emitter positions.

$$Y = f(X) = f(X(\alpha_i, \tau_i)). \tag{6.1}$$

An advantage of learning the target function from the training data is to realize how well the model generalizes to new data input. Therefore, generalization is necessary because the collected data sets work as a sample, which is incomplete and noisy. The task of a good machine learning model is to generalize well from the training data to any data from the problem domain.

The goal is to make a fingerprint that undergoes constant improvement via anchor emitter, which performs a recursive averaging and recalculation of the multipath fingerprint patterns. This hypothesis is based on the relationship between dependent and independent variables in TDOA, including NLOS identification.

Ray-tracing can play an essential role in the ML framework, not only in the scenario description but also in the extractions of multipath patterns. Therefore, the ray's description should be adequate to establish the model but not as precise that the model loses the generalization feature.

This effect is well-known in the literature as over-fitting and under-fitting of the data set, the most significant cause involved in the poor performance of machine learning algorithms [Xie et al., 2018]. The multipath based on the ray-tracing data set should present some typical channel characteristics. If the multipath description is good enough, the supervised machine learning algorithms can find the best approximation to the unknown mapping function for the output variables, given the input. Machine Learning and Kernel Methods can work as a supervised, unsupervised, or semi-supervised learning framework.

In this contribution, RT has been used for the initial data sets, and the CIR estimation of the reference or anchor nodes works as a "reinforcement" or supervised learning.

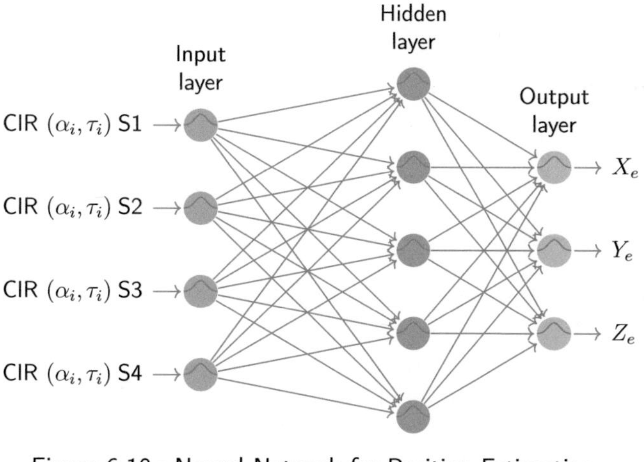


Figure 6.10.: Neural Network for Position Estimation.

The problem was designed to predict the emitter position, i.e., the X, Y, and Z coordinate using **linear regression** when the delay and amplitude of the five multipath rays were given. The network had ten inputs, i.e., five rays. Neural networks, linear regression, and the random forest algorithm were used to achieve the result. In our experiments, the random forest algorithm proved to be the best method among the three.

However, one of the problems encountered in this model was having less than five paths. It was necessary to assign these point 0s to give the same length input to the network. A set of hyper-parameters for tuning linear regressions [Neter et al., 1989] was used, which provided better results.

The proposed algorithm can be summarized in Algorithm 3.

**Algorithm 3:** TDoA with RT Fingerprints using NN

**input** : CBB signal in each Sensor $S_1$ to $S_4$
**output:** Estimated Emitter Position
**From a deployed TDOA Sytem:**;
· Record the signal, extract the CBB;
· Apply **SIMO FIR** or **MUSIC** in each CBB for 5 tap-delay estimation 5 Multipath Components, $\alpha_i, \tau_i$ $[CIR_1, CIR_2, CIR_3, CIR_4]$;
**Perform Ray Tracing Simulation of the Scenario with:** ;
· TDOA Sensor Positions ;
· Emitter position vector $\boldsymbol{x}$ ;
· Reference Emitter position vector $\boldsymbol{x_{REF}}$ ;
· Buildings descriptions.;
**With RT output collect Produce a CIR Fingerprints**;
5 $(\alpha_i, \Delta_{\tau_i})$ in each CBB;
**Build Neural Network:**;
· 40 Inputs: 5 Peaks $(\alpha_i, \tau_i)$ for each CBB;
· 8 Neurons in Hidden Layer;
· Weights and hyper-parameter refinement;
· pre-processing data set for Machine Learning Engine **Perform CIR for feature of Reference extraction:**;
· Record CBB from the Calibration;
· Perform SIMO FIR CIR for reference position;
**Enhance NN with the Augmented data sets:**;
**Perform the Position Estimation:**;
**Repeat the Estimation for a new "recording" windows:**;
**Perform the joint position estimation with ICI**
$P(\mu_{TDOA}, \sigma_{TDOA}) \cap P(\mu_{RT}, \sigma_{RT})$

In short, the proposed method combines CIR estimation,RT simulation, and machine learning to match the amplitude and delay information to a defined position. The method is similar to the RSSI [Zekavat and Buehrer, 2011] if the amplitude or the received power information is used. The fingerprints produced by five rays in four different receivers add more dimensions to the matching procedure.

### 6.1.4. RT Multipath Fingerprints using Random Forest

As described in the Chapter 5, there are several different implementations of machine learning; however, for this research, regressions and classification models based on Random Forest have been used. High non-linearities and the considerable number of different classes lead us to define our method using this approach.

The ideas presented in [de Sousa and Thomä, 2019a] and [de Sousa and Thomä, 2019b] can be revisited, which define our fourth contribution, to use a random forest model to match the RT simulations and feed the data sets.

The CIR from each RT is used to produce a Volume Cross-correlation from the pair 2-1,3-1 and 4-1. In our framework, the dataset is built by taking the possible emitter position $(X_e, Y_e, Z_e)$ and the peaks (amplitude and delay) in the VCC as the features to be used by the Random Forest classification engine. Moreover, the five maximum peaks in VCC are structured as inputs in the machine learning engine.

As we use a classification framework, the RF produces several decision trees randomly distributed using the features in the data set. The end of the classification process is one position defined as a label in the dataset formulation. The emitter position estimation is done with every "set" of data from the VCC "peaks." RF makes an approximation to find a position as a classification process $(f)$. This function should be able to map input variables $(\Delta_{21}, \Delta_{31}, \Delta_{41})$, to an output variable $(X_e, Y_e, Z_e)$, which is the emitter position.

$$[X_e, Y_e, Z_e] = f(\Delta_{21}, \Delta_{31}, \Delta_{41}). \tag{6.2}$$

The proposed procedure is summarized in Algorithm 4:

**Algorithm 4:** RF Fingerprint in TDOA Systems

**input** : CBB signal in each Sensor $S_1$ to $S_4$
**output:** Estimated Emitter Position
**From a deployed TDOA System:**;
· Record the signal, extract the CBB;
· Extract VCC TDOA with 5 Peaks for each TDOA $[\tau_{21}, \tau_{31}, \tau_{41}]$;
**Perform Ray Tracing Simulation of the Scenario with:** ;
· TDOA Sensor Positions ;
· Emitter position vector $\boldsymbol{x}$ ;
· Buildings descriptions.;
**With RT output collect Produce a VCC Fingerprints**;
5 $(\alpha_i, \Delta_{\tau_i})$ in each TDOA;
**Build a Classification Random Forest:**;
· 30 Features: 5 Peaks $(\alpha_i, \Delta_\tau i)$ for VCC TDOAs;
· 3 Outputs: Emitter Position $[X_e, Y_e, Z_e]^T$;
**Refine the dataset of RF:**;
· Record from the Calibration;
· Perform VCC CIR extract reference;
**VCC TDOAS from calibration points**;
· Perform "reinforcement";
· Adjust the data set of Random Forest Engine;
**Perform the Position Estimation**;
**Apply the Random Forest method and refined the data-set**;
**Repeat the algorithm from Step 4 with a new integration windows until the end of the recording signal.**;
**Perform the joint position estimation with** $P(\mu_{TDOA}, \sigma_{TDOA}) \cap P(\mu_{RT}, \sigma_{RT})$.

## 6.2. Processing the Datasets for Multipath Fingerprints

This section describes the creation of multipath information and its use in the validation scenario. It demonstrates how the grid points were defined and used to generate the RT fingerprints. Every point in the validation scenario is defined as one grid point in the machine learning framework.

As explained in Chapter 4, the RT simulation can offer a multipath exploitation scheme. Therefore, for each TDoA sensor, there is a characteristic CIR for each possible emitter-sensor connection. The section also describes how to connect the information produced by the reference emitters with the data sets produced by the RT.

### 6.2.1. RT datasets

Initially, a data set is produced using only RT simulation, where the sensor positions, the general description of the emitter target (power output, the central frequency used), and the environment obstacle description (buildings, trees, roads) are fed to the simulation.

This process produces a grid in the area that is defined in the ray-tracing resolution. Particular attention should be given to the fact that the grid size should not overload the computer processing scheme.

For our research, we have adopted a grid size of 3 meters for Ilmenau 2 and Uttenreuth 2; and a grid size of 1 meter for Ilmenau 1 and Uttenreuth 1 scenarios. Further details and a better description of each scenario will be discussed in the experimental evaluation section during the definition of each experiment used.

At this point, it is imperative to remember the collection region aspect of the RT simulation, explained in Section 4.2. There is a trade-off between the grid size resolution and the size of the collection point itself in the RT simulation. In an outdoor scenario, with 2km × 2km, the description of buildings, obstacles, and material characteristics also adopt some degree of approximation in RT itself. Details about these data are found in section (Please refer to section)

Figure 6.11 shows the assignment procedure adopted in the case of our data set formulation. Every point in our scenario is assigned to the grid center. As we are dealing with Euclidean distances in the calculation, the UTM coordinate system was adopted for the grid formation.

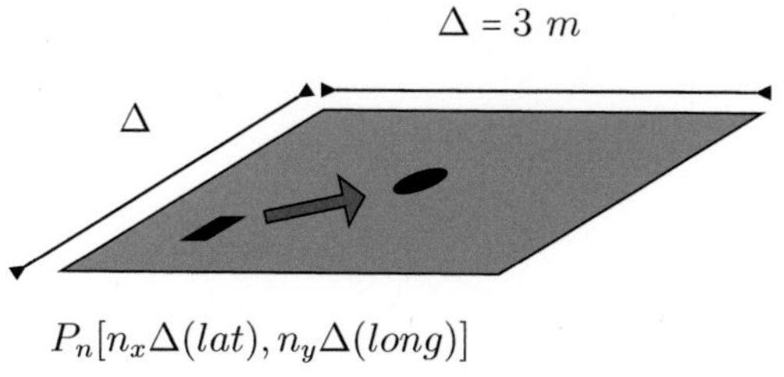


Figure 6.11.: Assignment of Points to the Grid Center.

Each discrete point has a unique position description, formed by the combination of the latitude and the longitude, as described in Section 4.5 Figure 6.14. A data set is created by matching position with the CIR fingerprints, acting like the features of the machine learning engine.Figure 6.12 illustrates the process of data set creation.

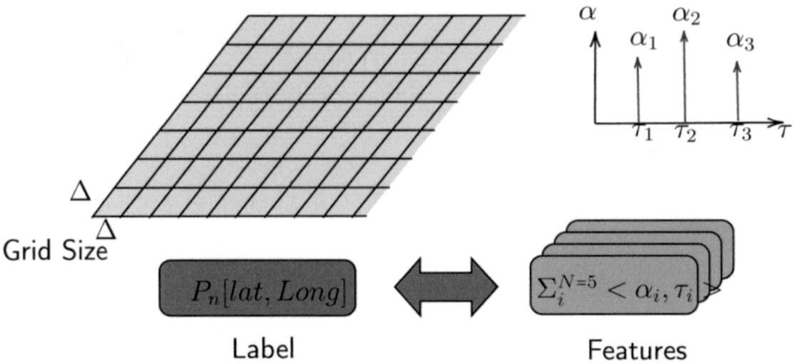


Figure 6.12.: Dataset Extraction from the RT.

The next step is to perform a data set pre-processing, which involves data-cleaning, data reduction, and data editing [Raschka, 2015]. In RT simulation, it might be possible to have some regions in the simulation domain that are not achieved by any rays; their values will be considered as zero.

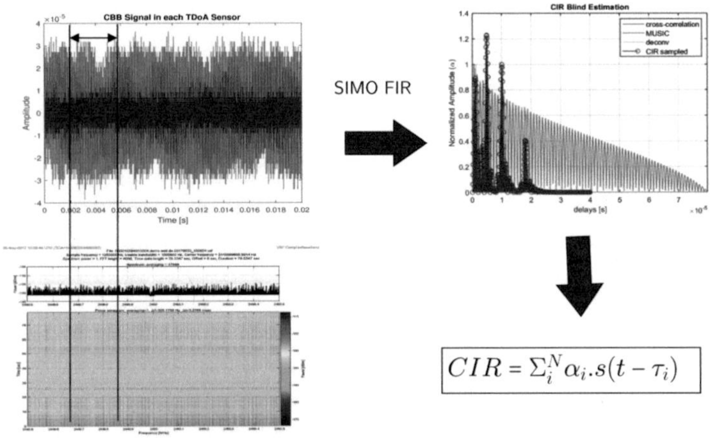


Figure 6.13.: CIR Estimation of the Reference Emitters.

The zero-values on a data set is a well-known machine learning problem because these null values can cause serious convergence problems when used in model training. To deal with this problem, we have assigned a minimum value for all the points in the scenario.

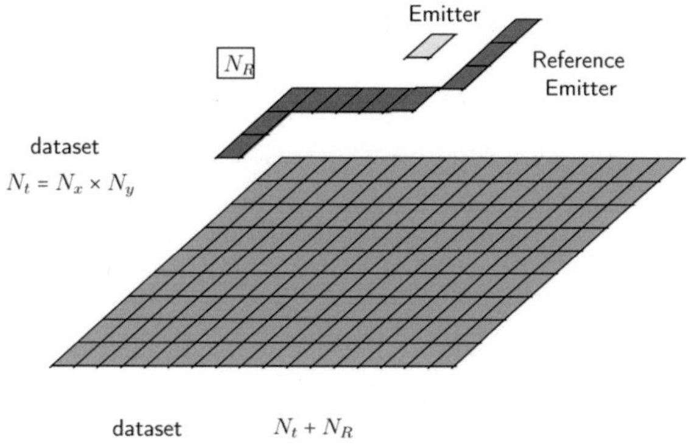


Figure 6.14.: Enhanced Datasets using Reference Emitters.

### 6.2.2. Reference Points Dataset

As outlined in Figure 6.1, during our validation procedure, the emitter has a track in different positions, where a CBB signal in each sensor is recorded.

All the actual emitter positions are assigned to defined points in the grid scheme in the grid scenario. Thus, the measurements from RT simulation and the real measurements coincide for these positions. One of the points in our approach is called an actual emitter, and the other points of the track are designated as reference points.

Using either the CBB from each sensor or the pure VCC produced in the correlation process, the multipath fingerprint characteristics of each reference point are extracted.

The CIR estimation is performed using the SIMO FIR approach, explained in Section 3.3, Section 3.4. We have further implemented the VCC of all the reference points, as suggested in Section 3.5. Figure 6.13 shows the CIR estimation process and the feature extraction to enhance the initial dataset produced by ray-tracing.

In short, now two datasets are created, one produced using only RT simulation and the other one with the real measurements.

## 6.3. Multipath Fingerprints Dataset Enhancement with Reference Emitters

This section explains a detailed procedure to correctly combine the information obtained by the RT simulation and the real measurements from the reference emitters. It is possible to apply the multipath fingerprints approach using only real measurements. However, it is sometimes impossible for operational reasons to have a data set produced using only the real measurements in an outdoor scenario.

Our approach for using RT data sets, enhanced with reference points, tries to overcome these limitations. Combining real data with simulated or synthetic ones is not a simple issue; it is challenging in machine learning research.

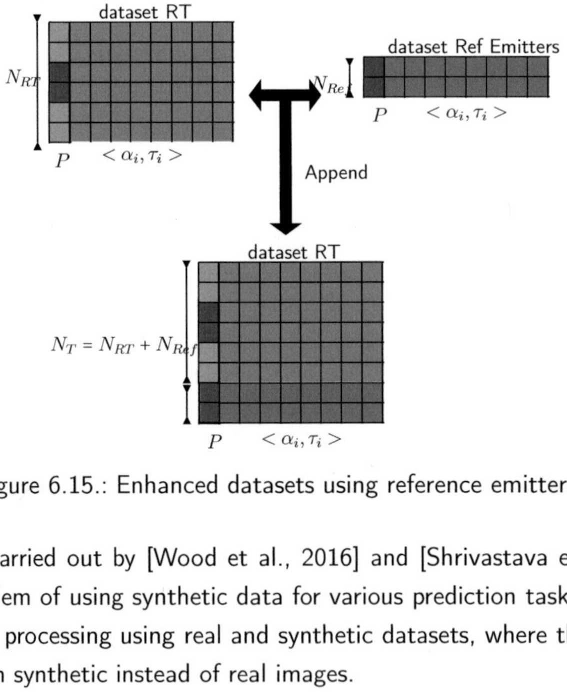


Figure 6.15.: Enhanced datasets using reference emitters.

The studies carried out by [Wood et al., 2016] and [Shrivastava et al., 2017] have explored the problem of using synthetic data for various prediction tasks. Their research focused on image processing using real and synthetic datasets, where they discussed the idea of training on synthetic instead of real images.

Figure 6.14 shows the dataset augmentation procedure to join the real measurements, and the RT fingerprints producing an enhanced dataset. Using only RT datasets may not achieve the desired performance due to a mismatch between synthetic and CIR distributions from the actual CBB produced in each sensor.

There are several well-established procedures to join the two datasets in machine learning problems [Raschka, 2015]. The most common is the *Merge* and the *Append* operations. In our research, we have used the aspects presented by [Wong et al., 2016], restricting our analysis to the *Append* approach that provides better performance, reducing over-fitting.

In our datasets, the labels or the classes are the positions in a defined grid, so it is considered a closed set, where each point in the domain has the properties that define it unequivocally. The two datasets assign the same possible points in the grid, so, in the end, we have the same points with information derived from both the synthetic RT simulation and the real measurements.

The **Append** operations add cases or observations to a dataset. Appending two datasets requires that both have variables with the same name. Once the [latitude-longitude] number uniquely identifies each point, no additional variables or names are created. The information is only added to the pre-existing variables.

There are several ways to implement the append operation in a dataset. [Raschka, 2015]. In the RT datasets, there are several points with zero information for the CIR. Some regions are not achieved by the simulation either the power energy is too low, or the stopping criteria are met, as discussed on Section 4.3.

Appending operation causes an increase in the dataset entries. It serves as a data augmentation procedure explained in Figure 6.14. Figure 6.15 shows the procedure schematically, where we can see that the dataset increases but with the same number of points. The extra and redundant information is used in the machine learning model training, either in the classification or in the regression model.

The overall effect of the dataset enhancement is to improve the classification model obtained. This procedure is following the findings presented in [Wood et al., 2016] and [Raschka, 2015]

In conclusion, the datasets produced either by real measurements or ray-tracing simulation are appended to create a mixed dataset. The final machine learning engine uses the enhanced dataset to create both the classification and regression models.

## 6.4. Concluding Remarks

This chapter explained the contributions presented in the previous sections and linked the research contributions with the research questions. The concept of multipath machine learning fingerprints was explored in-depth, along with implementation details.

The integration of different estimations considering the enhancement proposal as a hybrid method, with an unknown cross-correlation between the position estimations, is discussed. Therefore, we used ICI as a data fusion approach.

We also explained the three localization enhancement approaches derived. At first, we used RT simulation and a database of possible TDoA; then, we applied regression with Neural Network was discussed, and finally, classification adopting the Random Forests model. This chapter ends with the dataset generation description, emphasizing the approach to append real and synthetic RT datasets in the machine learning models.

With the issues described in Chapter 2, and the techniques for extracting the multipath features from the real measurements given in Chapter 3, we could apply the concepts introduced in Chapter 4 and build a machine learning engine, based on the random forest model, as explained in Chapter 5.

# 7. Performance Evaluation

In this chapter, we demonstrate the performance of the RT multipath fingerprinting method using machine learning to enhance Radio Frequency (RF) outdoor localization systems. We have applied it in two experimental scenarios with different emitters.

Although we have concentrated our research on TDoA systems, our idea could also be further extended to other localization techniques and different scenarios. The experimental setup used here was selected to demonstrate that it is possible to employ multipath to improve the accuracy of positioning systems in general.

We have used a mix between synthetic datasets produced by RT and actual measurements from calibration emitters. Consequently, we designed the experiments to fuse the multipath fingerprints, enhancing the overall localization performance. We picked two small cities in Germany for validation scenarios, Uttenreuth and Ilmenau.

Uttenreuth is located in the middle Franconian district of Erlangen-Höchstadt in Germany, and it has small buildings widely spaced, including old and traditional houses. Thus, it could be defined as a typical rural area propagation environment (as per [Series, 2015]).

Ilmenau is a small town in Thuringia, Germany, located approximately 33 km south of Erfurt. The houses, buildings, roads, and landscape are typical suburban areas with small Doppler effects and possible minor shadowing effects in the propagation signal caused mainly by moving vehicles.

The chapter is structured as follows: first, we describe the measurement setup, including emitters, TDoA system description, the signal used, and other mechanical and electrical factors. We have included in our analysis the Confidence Ellipsis (GDOP) and the CDF as evaluation criteria (as explained in Appendix A).

Section 2 explains the steps to generate the dataset representing the multipath position fingerprints, emphasizing how we deal with the mismatching between real and simulation values.

Section 3 shows the validation application in the enhancement in the Line Of Sight

(LOS) and NLOS situation. First, we explain the TDoA database, the neural network regression, and Random Forest classification, as discussed in [de Sousa and Thomä, ], [de Sousa and Thomä, 2018a] and [de Sousa and Thomä, 2018b]. In the last scenario, we show the fingerprinting method using multipath data produced by VCC function, as presented in [de Sousa and Thomä, 2019a] and [de Sousa and Thomä, 2019b].

In Section 4, we perform an in-depth analysis of the results, explaining the machine learning parameters and the ICI fusion used.

The chapter concludes with an in-depth analysis of the possibilities and limitations of the proposed framework, showing the findings of the research based on the testing application. We present the complete cycle of RT fingerprint multipath enhancement, including its machine learning implementation. In the end, we perform a connection between our findings and the research problem addressed, summarizing the key elements discussed.

## 7.1. Experimental Scenario Configuration

This section aims to describe the general aspects of the outdoor scenarios chosen to validate the RT multipath fingerprint enhancement approach. These measurements were conducted in the cities of Ilmenau and Uttenreuth to test the performance of the multipath fingerprint in outdoor scenarios. We explain each experiment, including equipment used, placement, the configuration of the emitter and the receivers, and the protocol established.

As described in Section 1.5, our method was developed and tested in small steps, adding refinements to achieve the final version. In the experimental scenario here, we have used the concepts introduced in Chapter 3, Chapter 4 and Chapter 5 to solve the problems described in Section 2.4.5.

The studies carried out by [Richter, 2005] and [Landmann, 2008] showed that the Specular Components (SC) are well defined and more significant than the Dense Multipath Component (DMC); therefore, the RT simulations used to build the multipath fingerprints can obtain a good match between simulation and real channel conditions.

Table 7.1 is based on [ITU-R, 2019] and shows the propagation environments characteristics of Ilmenau and Uttenreuth.

Table 7.1.: Environments and Propagation Characteristics following [ITU-R, 2019].

| Place | Enviroment Classification | Description and propagation characteristics |
|---|---|---|
| Utternreuth | Rural/ Residential | - Small house surrounded by extensive gardens;<br>- Single and double storefronts;<br>- Roads are generally two lanes wide with cars parked along sides;<br>- Heavy to light foliage possible;<br>- Motor traffic usually light (it is far from highways);<br>- Terrain is an influence due to the height (topography). |
| Ilmenau | Urban/ Suburban Lowrise | - Typified by wide streets;<br>- Building heights are generally low making diffraction over rooftop likely;<br>- Reflections and shadowing from moving vehicles can sometimes occur;<br>- Primary effects are long delays and small Doppler shift. |

After this basic description of the two cities, it is necessary to explain in detail the TDoA localization system used, the configuration and setup used for the radio-frequency emitter, and the machine learning framework used to build the multipath fingerprints datasets.

As shown in Figure 1.4, the RT multipath fingerprints framework operates as several connecting gears, where the CIR estimation using VCC or SIMO FIR, together with the RT simulations feed a machine learning engine to enhance the TDoA localization performance. We show the interconnection of these ideas to produce a set of estimations that are later fused with the real measurements to improve the overall result.

## 7.2. Generation of the Multipath Fingerprints using RT

In this section, we present the RT simulations done in each scenario previously described. The desired multipath fingerprinting scheme is achieved as discussed in Section 4.5, and the details for grid assignment are provided following the procedure presented in Section 6.2.1. (The grid size is an optimization problem because it depends not only on the scenario description details but also on the definition of the capture region for the RT simulation.)

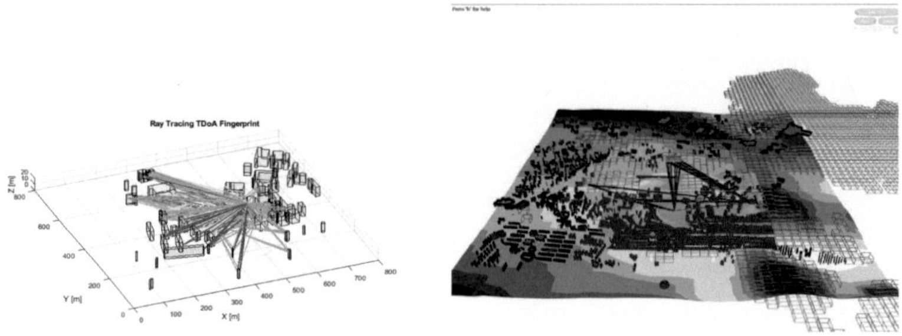


(a) RT from Bologna, [Fuschini et al., 2015].

(b) with AWE [Hoppe et al., 2017].

Figure 7.1.: Dataset extraction using RT in Ilmenau.

The buildings and obstacles in the scenario were extracted from Open Street Map databases [Bennett, 2010]. Due to this, some refined details are omitted from the simulations. The magnetic permeability, electrical permittivity, and conductivity was assumed constant.

The emitter specifications and the sensor position for each measurement scenarios are described in Table 7.4, Table 7.5, Table 7.6 and Table 7.7. The technical specification for each sensor is displayed in Table 7.8.

We have used two different tools for RT simulations in creating the multipath fingerprints: an in-house RT tool from Bologna University, [Fuschini et al., 2015], shown in Figure 7.1a and a commercial solution based on AWE WinProp [Hoppe et al., 2017], shown in Figure 7.1b.

The RT results needed to be post-processed to build the multipath datasets. In the case of CBB extraction with SIMO FIR, the fingerprints are directly used, taking five rays from each TDoA sensor deployed in each point of the grid. In this case, the dataset used $40$ features; meaning the $\alpha, \tau$ for each of the five rays in each sensor.

In the VCC approach, it was necessary to produce a synthetic cross-correlation operation. After this step, it was required to extract the five peaks in the simulated VCC signal.

As discussed in Section 2.4, the TDoA sensor connection link imposes a trade-off between the desired resolution and the feasible bandwidth. Therefore, the ability to extract useful multipath information in the signal processing tools required up-sampling strategies and statistical processing approach.

The association between RT and real measurements was the bottleneck of the the method itself due to mismatching and noise effects that corrupts the dataset values.

Some reference points and parts of the simulation domain lack information in the dataset creation procedure, called mismatching. This wrong information is assumed to be noise.

Consequently, we use the measures that we could extract even with significant differences in the datasets. The procedure of *data augmentation* discussed in Section 5.7 was taken into account in the dataset formulation. We have adopted the values $1 \times 1$, $3 \times 3$ and $5 \times 5$ meters for the grid definition for the fingerprint generation. Figure 7.19 shows the RT dataset extraction procedure use in Uttenreuth campaign.

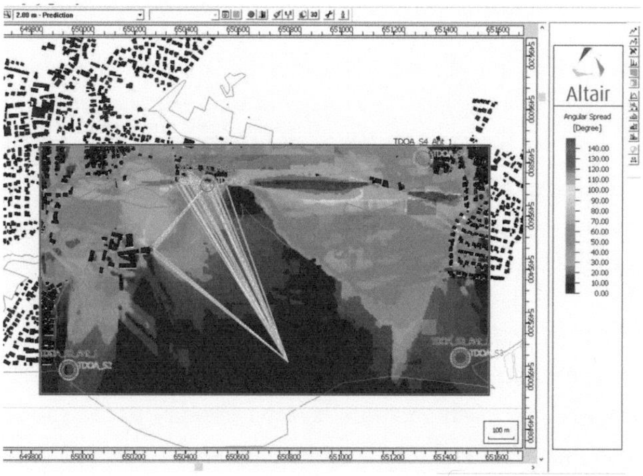


Figure 7.2.: Dataset extraction in Uttenreuth using AWE RT, [Hoppe et al., 2017].

The machine learning parameters used in our validation scenario is a critical aspect in the Random Forest implementation to refine the model in our multipath fingerprinting structure. Although our method could be applied using different machine learning implementations, we have restricted this research to the classification and regression approaches from Random Forest models.

The aspects presented in Section 5.4 and [Louppe, 2014] have driven us to choose these methods in our outdoor propagation scenarios. Defining the machine learning parameters is a challenging topic because it is not straightforward to determine which one has more important than the others.

The settings can either increase the model's predictive power or make it easier to train the model. All the parameters have their function that precisely affects the overall result. Following [Géron, 2019], the changing of one hyper-parameters can affect the efficiency of the complete prediction. There are some more relevant than others, but this relevance will depend on the nature of the task and the datasets.

There are helpful guidelines to start the model training [Géron, 2019], but the real evaluation can only be done during the model application. We can only assure good results when we set the model to work in real problems.

As an ensemble method, the Random Forest model combines predictions of several estimators to be applied in data that was not in training. The combined estimator is usually better than any single base estimator, and its variance is reduced.

In our application scenario, it means that the method should be able to cope with the differences between RT simulations and real measurements. As presented in Section 5.5, our datasets can have $40$ features, in the case of CIR multipath, or $30$, in the case of VCC peaks. We compared the actual measurements and the final result achieved, and we have concentrated our refinements on the following parameters: **number of estimators**, **maximum number of features** and the **maximal depth** in the trees.

The **number of estimators** is the number of trees in the *forest created*, in theory, if we use a considerable number of trees. We get better prediction results, but it demands more computing power. For each scenario, it was necessary to use the analysis similar to one presented in Section 5.8. In a classification approach, it is recommended to use the number of estimators as the square root of the maximum number of features.

The **maximum number of features** corresponds to the multipath values of the considered fingerprint (CIR or VCC). We have decided to use all features in our training model; it would be possible to use the Principal Component Analysis, as suggested by [Géron, 2019] and [Raschka, 2015], for instance, dimensionality reduction.

The **maximum depth** represents how deep each decision tree goes in the training procedure. When the tree is more in-depth, it is possible to have more splits, capturing more information about the data. In the multipath fingerprints, the decision deepness is related to the grade of refinement to assign a collection of points that are close to the target.

There is a trade-off between desirable precision in the localization analysis and the tendency to over-fit the model. We see that RT multipath datasets can overfit large depth values. It means that a tree can correctly predict the training data. However, it fails to generalize the findings for new multipath information.

Table 7.2 summarizes the key parameters defined in the validation scenario. The other machine learning parameters were adopted with the *default* recommendations.

Table 7.2.: Random Forest Parameters Settings.

| Random Forest Parameters | Regression | | Classification | | Obs |
|---|---|---|---|---|---|
| | CIR | VCC | CIR | VCC | |
| **Nr estimators** | 13 | 13 | 6 | 6 | Nr of Decision Trees |
| **Max nr of Features** | 40 | 30 | 40 | 30 | Multipath Features |
| **Max Depth** | 20 | 30 | 20 | 30 | Levels in Decision Trees |

In short, we refined these three parameters based on the field test results and the integration with the simulated RT simulations. In our non-collaborative localization procedure, this parameter adjustment could be performed in the reference emitter features.

### 7.2.1. Emitter Configuration

During our validation scenario, we used three different types of transmitters. The first signal was from a typical UAV control and video link, based on WiFi technologies and operating at 2.4 GHz. We have used the DJI Phantom 3 [DJI, 2020] (a consumer drone with an onboard camera).

Table 7.3.: DJI Phantom Specs technology controls used, based on [Abdulkabir, 2019].

| Type | Frequency | Modulation | Technology |
|---|---|---|---|
| controls | 2.4Ghz/5.8Ghz | FHSS/DSSS | FASST/Lightbridge |
| video | 2.4Ghz | OFDM | Lightbridge/Wi-Fi |

As explained in [Abdulkabir, 2019], drones generally operate radio-controlled devices, usually with center-frequency in 2.4 or 5 GHz, a bandwidth of 20, 40 or 80 MHz, and an

Effective Isotropic Radiated Power (EIRP) around 20 dBm. Table 7.3 shows the common control and video channel specifications used by drones. The Figure 7.3 shows CBB drone signal received by the TDoA sensor .

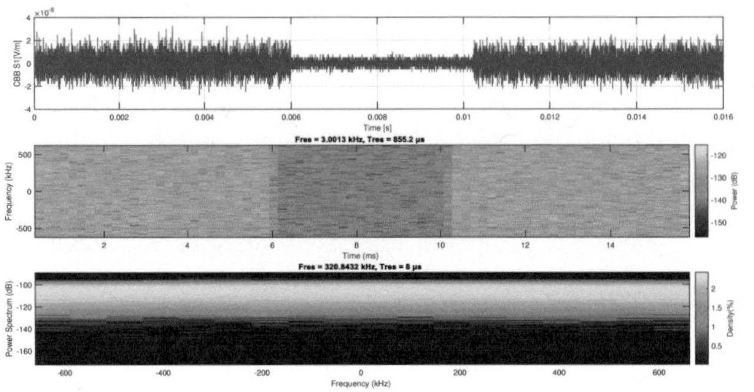

Figure 7.3.: CBB Received in TDoA Sensor 1.

The second type of emitter was a vector signal generator [Schwarz, 2020] directly installed in a car, with all proper electrical and physical adaptation to feed and transmit the signal as a mobile target. We have used the internal processing unit to set a pseudorandom binary sequence (PRBS) signal following the guidelines given in [ITU-T, 2000].

It is a periodic, deterministic signal created with a series of digital ones and zeros, using a random pattern within the repetition period. It was generated and transmitted with an EIRP of 13 dBm.

Figure 7.4 shows the mobile setup with the vector generator and one whip antenna on the vehicle rooftop. During the vehicle preparation, several mechanical and electrical adaptations were carried out to adjust the transmitter for continuous operation for four to five hours of measurements. It was necessary to install inverters and an extra power supply.

The final type of emitter was a signal generated by a USRP [NI, 2014] assembled both with and without power amplifier, also adapted to operate in a car as a mobile emitter. Figure 7.5 shows the power spectral density of the OFDM signal generated by the USRP.

Figure 7.6 shows the mobile configuration of the USRP. Figure 7.7 shows the USRP emitter without amplification installed in a wheeled cart. The OFDM signal generated by

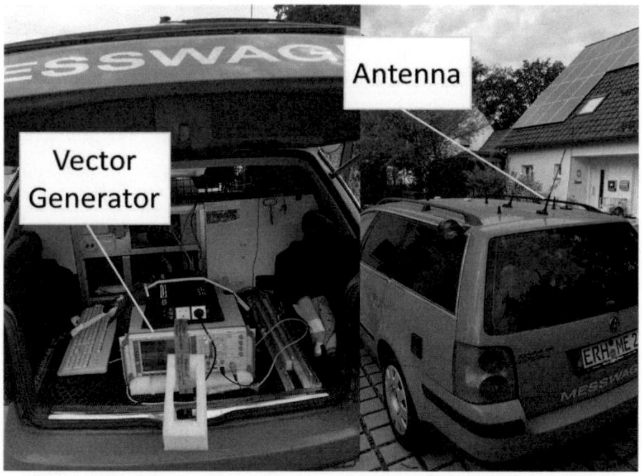


Figure 7.4.: Mobile emitter configuration using R&S SMU 200A.

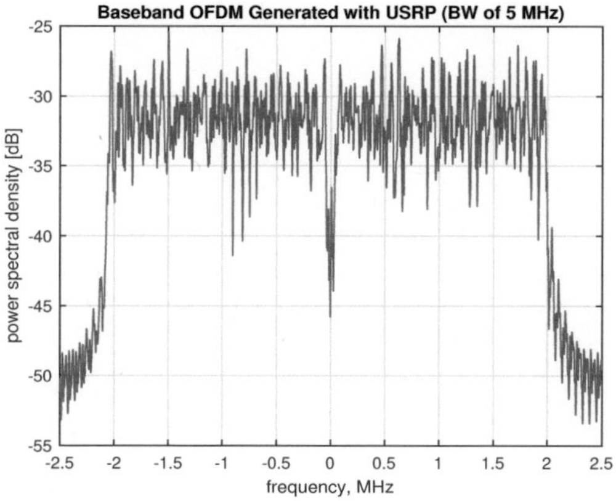


Figure 7.5.: OFDM Signal generated by USRP.

the USRP in its time and frequency domain is presented in Figure 7.8.

Table 7.4 shows a summary of the specifications of the different emitters used in our measurements.

We performed three sets of measurements in two cities. In Ilmenau, the emitter followed two itineraries; a small one is shown in Figure 7.15, and the other measurement set used a more extensive setup, so the mobile transmitter followed the route showed in Figure 7.16.

Figure 7.17 shows the validation scenario in Uttenreuth, including the emitter, sensors,

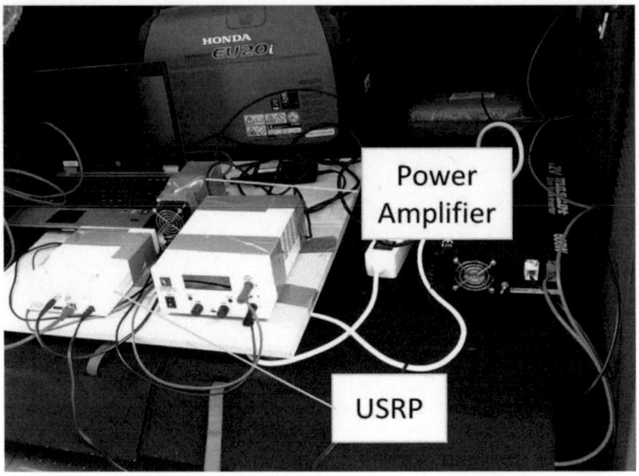


Figure 7.6.: USRP with power amplifier in mobile emitter configuration.

Figure 7.7.: USRP installed in a wheeled cart without amplifier.

Table 7.4.: Summary of the emitter specifications used in our experiments.

| Emitter | Central Frequency | Bandwidth | Modulation | Power EIRP |
|---|---|---|---|---|
| DJI UAV | 2.53 GHz | 20 MHz | FHSS/DSSS OFDM | 20 dBm |
| R&S CMU 200A | 433 MHz | 1 MHz | PRBS (15) | 13 dBm |
| USRP | 2.53 GHz | 5 MHz | OFDM | 10 dBm |
| USRP w Power Amplifier | 2.53 Ghz | 5 MHz | OFDM | 17 dBm |

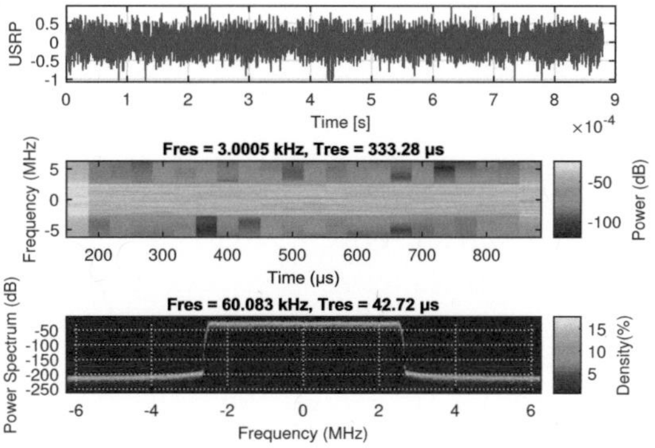


Figure 7.8.: OFDM Signal Tx by USRP.

and itinerary. In the first one, the UAV emitter was stopped in the same place while the **R&S CMU200A** emitter was set to a mobile configuration.

In total, we used, in our experiments, two distinct frequencies and three different emitters in various combinations. This diversity of emitters helped us to extract a more generalized result on multipath fingerprints, following the guidelines established in [ITU-T, 2000].

### 7.2.2. TDoA System Setup

We performed two sets of measurements, one in Ilmenau and one in Uttenreuth. Figure 7.15, Figure 7.16, and Figure 7.17 show the emitter itinerary in each validation scenario, the general view of each layout, and the TDoA sensor deployments in Ilmenau. The first setup was done in a small region, around $300 \times 300$ meters, while the second one was done in $3 \times 3$ kilometers.

The system used in our experiments was the TDoA KCS systems from Saab Medav Technologies [SMT, 2020]. The system can handle hybrid sensors, with different types of sensor nodes with Direction Finding (DF) capabilities to perform TDoA or Angle Of Arrival (AoA) localization.

The basic configuration adopts a node as the central station that supervises and controls the location missions at remote sites. The sensors are grouped according to the tasks or position characteristics. TDoA sensors can be installed in different platforms, either static (tower) or mobile (e.g., vehicle installation).

The KCS works as a system-level solution incorporating a sensor network of several sites with a flexible architecture that can be modified to adapt to new situations.

Table 7.5.: Position of TDoA sensor and emitter in Ilmenau small setup.

| | Lat [UTM] | Long [UTM] |
|---|---|---|
| Sensor 1 | 5616264.65 m N | 636878.89 m E |
| Sensor 2 | 5616170.91 m N | 636883.23 m E |
| Sensor 3 | 5616232.62 m N | 636793.09 m E |
| Sensor 4 | 5616373.57 m N | 636831.67 m E |
| Emitter (LOS) | 5616290.24 m N | 636798.40 m E |

Table 7.5 and Table 7.6 display the TDoA sensor and emitter positions for the Ilmenau measurements, while Table 7.7 shows the Uttenreuth measurements.

Table 7.6.: Position of TDoA sensors and emitter in Ilmenau, wider setup.

| | Lat [UTM] | Long [UTM] |
|---|---|---|
| Sensor 1 | 5616291.39 m N | 637376.24 m E |
| Sensor 2 | 5616248.33 m N | 635168.65 m E |
| Sensor 3 | 5614043.72 m N | 637042.57 m E |
| Sensor 4 | 5617640.20 m N | 636866.54 m E |
| Emitter (NLOS) | 5616282.46 m N | 637025.96 m E |

Table 7.7.: Position of TDoA sensors and emitter in Uttenreuth.

| | Lat [UTM] | Long [UTM] |
|---|---|---|
| Sensor 1 | 5495752.03 m N | 650393.25 m E |
| Sensor 2 | 5495834.98 m N | 650825.49 m E |
| Sensor 3 | 5494735.56 m N | 650711.63 m E |
| Sensor 4 | 5495001.41 m N | 649933.86 m E |
| Emitter (NLOS) | 5495286.19 m N | 650472.43 m E |

The sensors are connected via LTE links (or GPRS/EDGE/UMTS), and the measurements are collected in the central station where the localization analysis is performed. In the setup used in our experiments, we have configured the system to operate with four TDoA nodes; the central station has a DF antenna.

Although the DF functionalities with AoA capability was available, we did not use it. The system uses the Allaris DF-301 antenna from [Alaris, 2020] for direction finding, for the hybrid AoA/TDoA central station, as shown in Figure 7.9.

As described in Section 2.4, there is a trade-off between the bandwidth desired and the available communication links connecting the sensors, as massive recording bandwidth can lead to impractical real-time operation.

Figure 7.9.: Alaris Antennas DF-301, from [Alaris, 2020].

Table 7.8.: IZT Specification, from [IZTGmbh, 2020].

| Technical Specification (Summary) | | |
|---|---|---|
| Frequency Range | V\UHF | 20 MHz - 3 GHz |
| Input Sensitivity | V\UHF<br>@ S/N= 10 dB | -114 dBm @ 3 kHz<br>-105 dBm @ 25 kHz<br>-92 dBm @ 500 kHz |
| Input IP3 | V\UHF | +24 dBm, typical (low distortion)<br>+13 dBm, typical (normal mode) |
| Noise Figure | V\UHF | 10 dB, typical (low distortion)<br>15 dB, typical (normal mode) |
| Bandwidth | V\UHF | 6.25 kHz to 24 MHz |

Each sensor node is a wide-band digital receiver that is able to detect, down-convert, sample, and send the CBB snapshots with timestamp references to the central station, as described in Figure 2.3 and Figure 2.4 in Section 2.4. The KCS was set to operate with the receiver family from IZT, shown in Figure 7.10, [IZTGmbh, 2020]. Table 7.8 shows the principal technical specification.

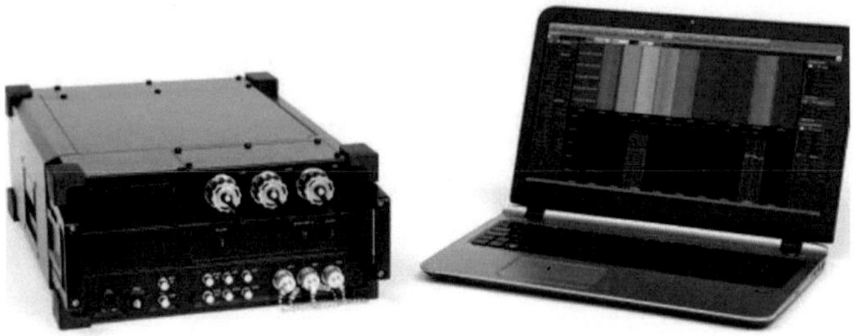

Figure 7.10.: TDoA Sensor node, from [IZTGmbh, 2020].

The synchronization of the KCS TDoA uses Global Positioning System (GPS) timestamps to mark the CBB snapshots, in a procedure already explained in Section 2.4.1. The antenna used in each node to record the RF target signal was chosen to cope with the frequency band used. The antenna was properly matched with the input impedance of the TDoA sensor.

Figure 7.11.: TDoA Node setup - ruggedized solution.

Figure 7.11 shows a TDoA sensor node installed, with the external GPS and the whip antenna. This station already has one internal processing unit, making it unnecessary to install an external one.

We also used a variant of TDoA sensor called IZT 3301. Both radios gave a similar performance, with the advantage that [IZTGmbh, 2020] is a lighter and more ruggedized solution that was suitable for mobile and rapid deployment operation. Figure 7.12 shows the variant solution for the sensor node.

Additionally, we also used the fixed facilities of SMT as a central node during the measurement campaign performed in Uttereuth.

The master station with DF capabilities is shown in Figure 7.13; it can record five CBB channels and provide extra information for our multipath fingerprints; Figure 7.14 shows the fixed antenna DF-A0029 [Alaris, 2020] used in Uttenreuth.

In short, in this section, we have explained the validation scenario, including a general description of each location and the critical specifications of the TDoA localization

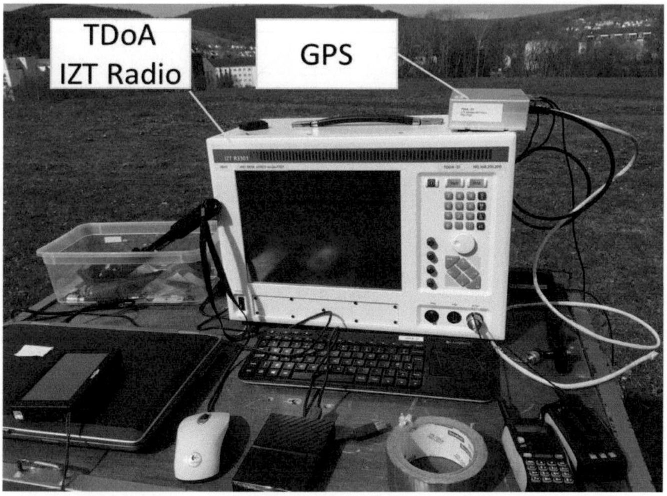


Figure 7.12.: TDoA Node setup with IZT 3301.

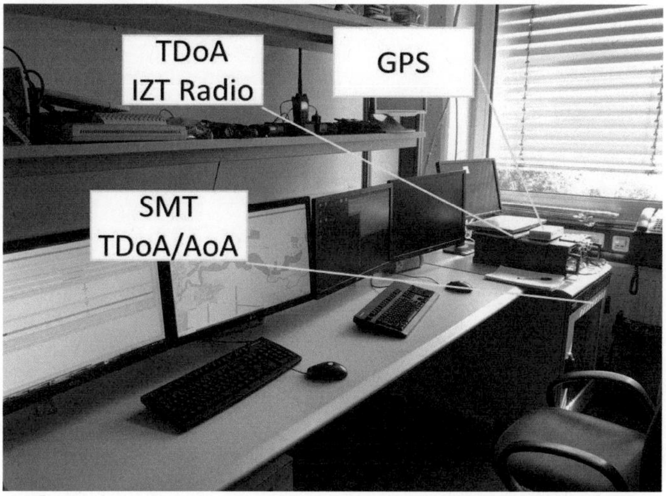


Figure 7.13.: Central station AoA TDoA installed in SMT facilities.

system used. We described the several emitters used and their signal characteristics, including power output, modulation type, and bandwidth. Finally, we reported the physical and electrical arrangements done in the sensor node and in the car to perform the measurements.

Figure 7.14.: Antenna DF central station in SMT facilities.

Figure 7.15.: Deployment in Ilmenau, UAV and OFDM Emitters, First Campaign.

## 7.3. Multipath Fingerprints Enhancement in TDoA Localization

This section shows the experiments carried out in the scenario described using the multipath fingerprints approach, joining real measurements and RT simulations in a unified machine learning structure.

The proper setup of the TDoA system, the emitters, the RT simulations and the signal processing tools are established to support our experimental validation. Table 7.9

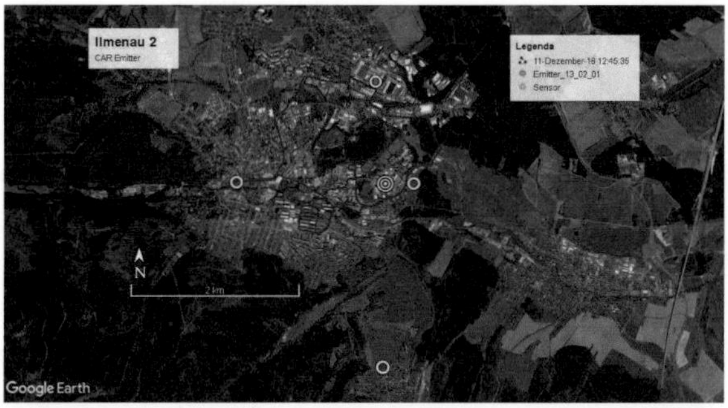


Figure 7.16.: Deployment in Ilmenau, OFDM Car Emitter, Second Campaign.

Figure 7.17.: Deployment in Uttenreuth, R&S CMU 200A Emitter.

highlights the main aspects. The setup was chosen to support the findings obtained to answer the questions posed in Section 1.3.

Table 7.9.: Settings for the validation experiments.

| Scenario | Emitter | TDoA Settings | RT Settings | Random Forest Settings | Fusion Strategy | Obs |
|---|---|---|---|---|---|---|
| **Ilmenau 1** | UAV<br>Table 7.3 | fc:2.53 GHz<br>BW:1 MHz<br>Only TDoA<br>Section 7.2.2 | grid:$1 \times 1$m<br>only buildings | none | Fingerprints<br>Intersection<br>Figure 6.4 | Database<br>Tdoa Fingerprints<br>Table 7.1<br>[de Sousa and Thomä, 2018e] |
| **Ilmenau 2** | OFDM CAR<br>Figure 7.15 | fc:2.53 GHz<br>BW:1 MHz<br>DF available | grid:$5 \times 5$m<br>(only buildings) | Regression<br>Classification<br>Section 6.1 | ICI<br>Appendix C | VCC Fingerprints<br>Reference Emitter<br>[de Sousa and Thomä, 2019a] |
| **Uttenreuth** | PRBS(15) CAR<br>Figure 7.6 | fc:433 MHz<br>BW:1 MHz<br>DF available | grid:$3 \times 3$m<br>(trees included) | RF Regression<br>RF Classification | ICI | VCC SIMO FIR<br>Reference Emitter<br>[de Sousa and Thomä, 2018b] |

Initially, we have applied the multipath fingerprints database to enhance a localization in a small-scale outdoor situation, using only the RT. The emitter was a DJI UAV, and the TDoA sensors were deployed in a $200 \times 200$ m space.

The second campaign was carried out also in Ilmenau, but with the TDoA sensor deployed in a $3 \times 3$ km area, using an OFDM emitter installed in a car. We applied the VCC to extract the multipath patterns information from the reference emitters. We employed Random Forest as a machine learning method in both setups, regression, and classification.

The last validation scenario was conducted in Uttenreuth, with the TDoA sensors in an area of $1 \times 1$ km, using the PRBS(15) sequence and a signal generator installed in a car as an emitter. We have applied the SIMO FIR approach for CIR extraction, and the classification and regression Random Forest was also used. We selected the GDOP and CDF for evaluating the results, as described in Appendix A.

The following subsections present the details and the result of each validation procedure in the different setups adopted.

## 7.3.1. Multipath RT Fingerprints Database

The first test was with a UAV acting as the emitter in a NLOS situation. The NLOS condition has produced substantial variations in the TDoA. As discussed in Section 2.4.5, during the processing, a traditional localization system, when evaluating the cross-correlations, would filter out those with quality lower than $50\%$, assuming that they are outliers.

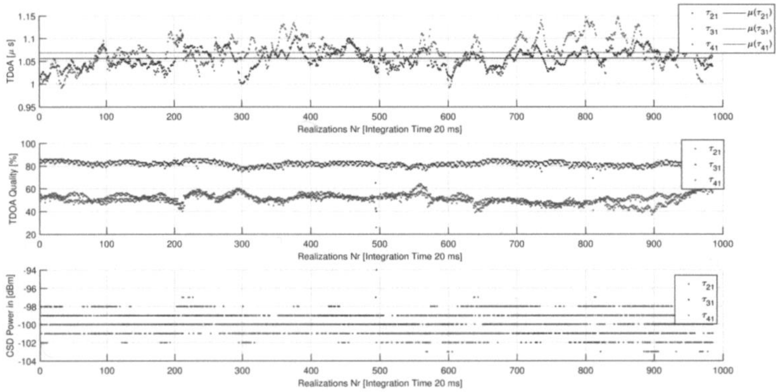


Figure 7.18.: Noise effects over Euclidean distance error for Random Forest model.

The measured time-differences oscillate from $\mu - \sigma$ to $\mu + \sigma$. We used this range together with RT simulations to produce an enhanced database with the ICI fusion of the estimates.

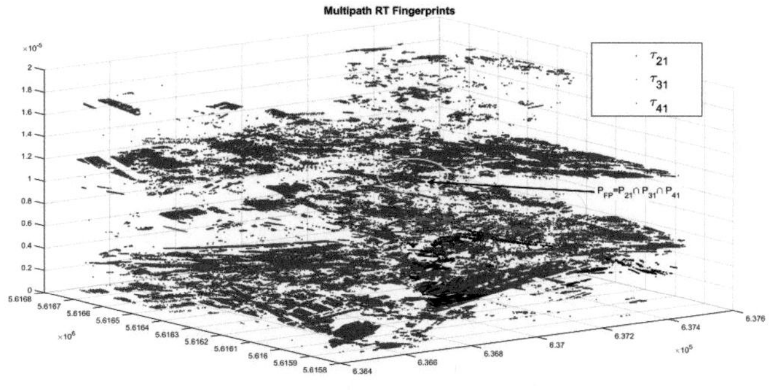


Figure 7.19.: RT Multipath Database. Lateral View.

Figure 7.18 shows the variations in the quality of the TDoA measurements in the testing scenario. The estimations were performed in several realizations, using 20 $ms$ as integration time. Looking at the TDoA results, the quality, and the power, we can observe that the moments with low quality and low power have substantial variations in TDoA estimations.

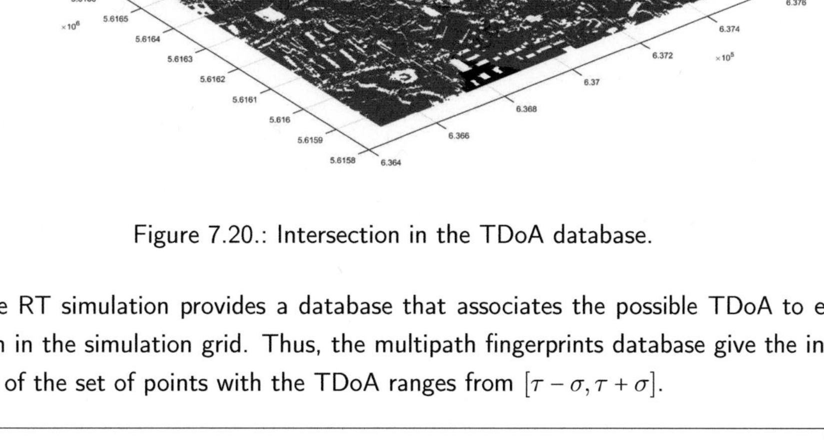


Figure 7.20.: Intersection in the TDoA database.

The RT simulation provides a database that associates the possible TDoA to each position in the simulation grid. Thus, the multipath fingerprints database give the information of the set of points with the TDoA ranges from $[\tau - \sigma, \tau + \sigma]$.

Figure 7.19 and Figure 7.20 shows the lateral and top view of the RT simulation, respectively. With a grid search, it is possible $\tau_{ij}$ to match the range of values with a defined area that represents the intersection, where it is likely to find the emitter.

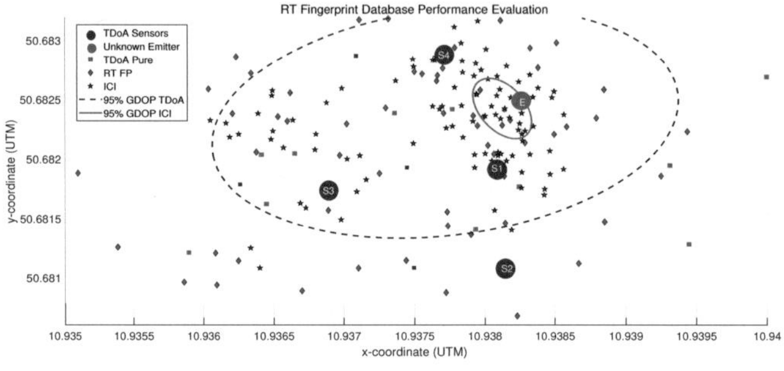


Figure 7.21.: RT Fingerprints enhancement in a TDoA.

Table 7.10 also shows improvements in performance. The ground-truth TDoA are calculated using the sensors and emitter actual position and the NLOS situation causes an error in the time-differences estimations, shown in terms of mean and variance $(\mu, \sigma)$. The joint estimation is carried out by fusing the multipath database and the real measurements.

Table 7.10.: Improvement in TDOA Position Estimation.

| TDOA | Ground Truth GPS | Measurements | | Proposed Method | Error Reduction (Improvement) |
|---|---|---|---|---|---|
| | | Mean ($\mu s$) | Std Deviation ($\mu s$) | ($\mu s$) | |
| $\tau_{21}$ | 0.333 | 0.114 | 0.0341 | 0.344 | 94% |
| $\tau_{31}$ | 0.27 | 1.76 | 0.0303 | 0.266 | 96% |
| $\tau_{41}$ | 0.266 | 0.15 | 0.0157 | 0.238 | 79% |

The GDOP of the fused estimations using ICI showed an enhancement in the overall position results, as indicated in Figure 7.21.

Figure 7.22 shows the error map produced by the RT fingerprints database. In this figure, estimation is performed for each point of the grid using the RT only. We can see on the map that the Euclidean error is big.

Figure 7.23 shows the CDF performance comparison of the enhanced TDoA system using the multipath RT fingerprints and the traditional one. The localization results

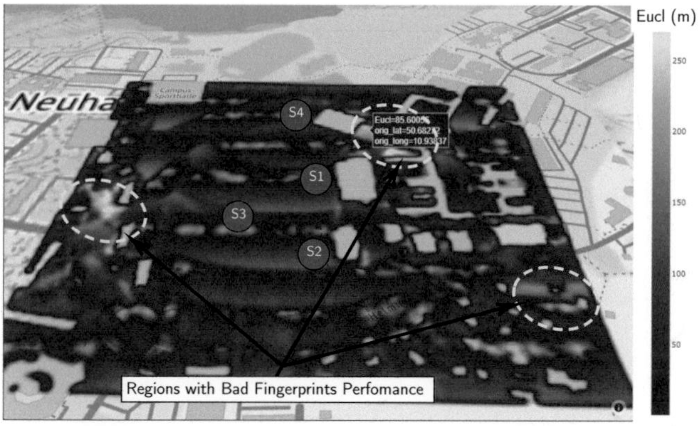


Figure 7.22.: Error map of the RT Multipath Fingerprints Enhanced.

were averaged and collected during 2 minutes of recording in the scenario described in Table 7.5.

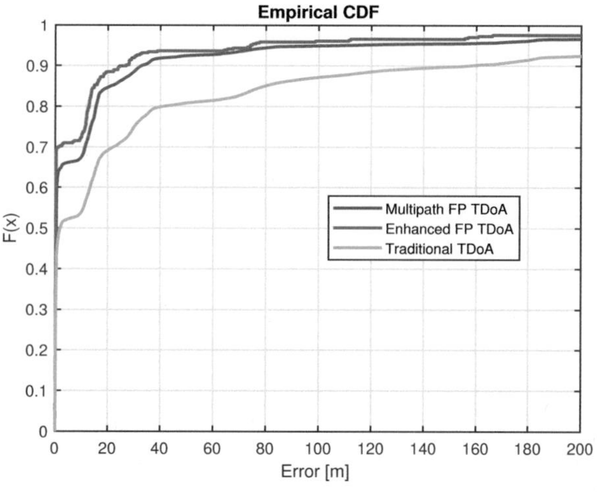


Figure 7.23.: CDF of RT Fingerprints enhancement in a TDoA

The first approach to the method was built using only one ray of simulation. Still, the promising results gave us the confidence to proceed with the idea by adding more radios to the simulation loop and using more elaborate technologies to construct the multipath fingerprint.

## 7.3.2. Random Forest Multipath Fingerprints with VCC

In this section, we describe the validation done in the outdoor environment. Still in the city of Ilmenau, this time, we unfolded the localization system on a broader scenario and used a transmitter attached to a car. After the promising results of using RT to improve the performance of the outdoor location, we took an extra step to enhance the method.

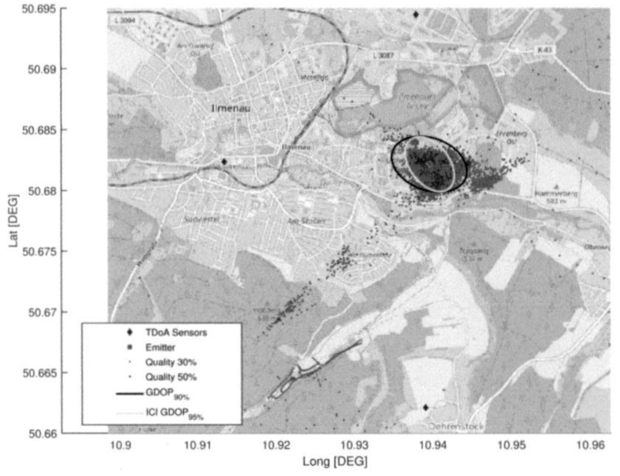


Figure 7.24.: TDoA including low quality Results ($Q > 30\%$)

At first, we presented the performance of a traditional location system in the same environment. Bearing in mind that the significant contribution from the multipath effects, the localization system cannot filter out all inadequate quality measures. We can see in Figure 7.24 emitter and sensor positions as weel the TDoA results of the $30\%$ and $50\%$ qualities. We have also added the GDOP of the pure localization systems and the enhancement achieved using ICI fusion with TDoA multipath fingerprints database.

In this application scenario, we used calibration emitters to generate a more robust dataset in conjunction with the simulations. We also added the machine learning techniques presented in Chapter 5, both in the classifiers model and the linear regression model.

The signal processing technique adopted was the VCC described in Section 3.5, where all the paths of the emitter were stored for further processing.

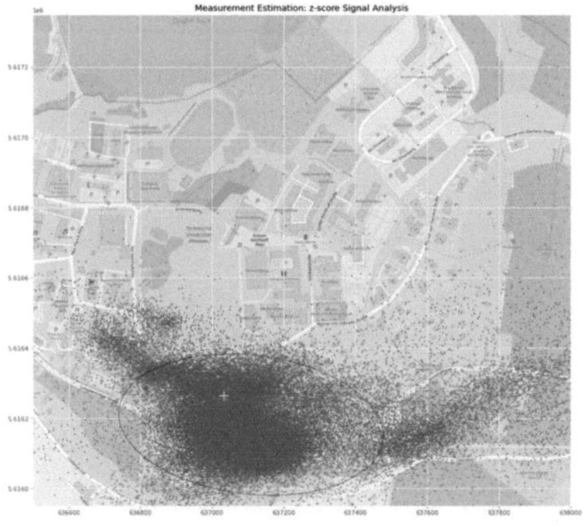


Figure 7.25.: TDoA System GDOP.

One specific point was chosen as the emitter; only signal characteristics were used for position processing at this location.

While a traditional system filters quality measures in the cross-correlation function below $50\%$, we have established that we would take into account measures down to a quality factor of $30\%$.

This procedure was necessary in order not to filter the NLOS effects that are essential in the characterization of the multipath in the environment.

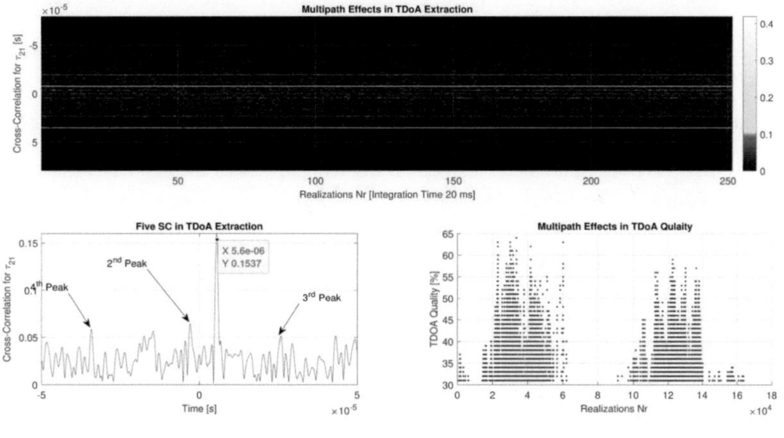


Figure 7.26.: VCC extraction from the reference emitters.

Figure 7.25 show the initial GDOP localization performance of the traditional TDoA. The emitter is indicated as a yellow "plus" mark, and the green dots are the estimation of TDoA system. We can see that the NLOS condition reduces accuracy.

Not all signals received from calibration transmitters were helpful in the dataset. Therefore, it was necessary to filter and choose those that could be used as reference signals. At some points in the testing environment, the emission was not received, and even if it was the multipath contribution, it was too random to be used in our method.

Figure 7.26 presents the VCC extraction procedure. As indicated, sometimes it is difficult to identify the other peaks in the cross-correlation function; in these cases, value zero is assigned to the dataset. As explained in Section 3.5, the position of the other peaks does not change considerably when we perform several realizations. The other peaks are related to the multipath composition of the CIR that connects the emitter to a sensor-pair. The main specular components can be retrieved from the other peaks in VCC processing.

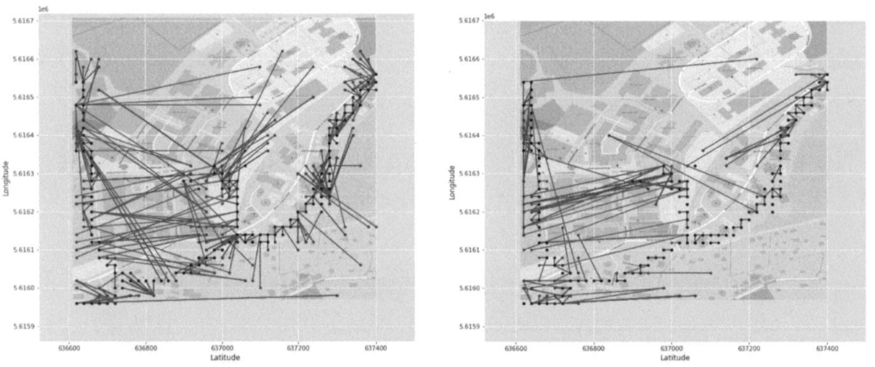

(a) Dataset created using only RT simulations. (b) Enhanced Dataset (RT and Reference Emitters).

Figure 7.28.: Machine Learning Random Forest Classification.

Figure 7.28 shows the result of the localization fingerprint scheme. The blue dots are the ground-truth positions of the reference emitters, and the green dots are the estimates made by the Random Forest. The red lines connecting the two points provide a visual evaluation of the error (Euclidean distance) between the reference point and the estimated one. The left side shows the performance using only the dataset produced with RT. On the right side, we notice the estimation improvement when the measurements of the reference emitter are added to the dataset.

A usual problem in machine learning is the identification of the correlation among the features. Our fingerprints strategy refers to how different each feature set that defines a point is from the others.

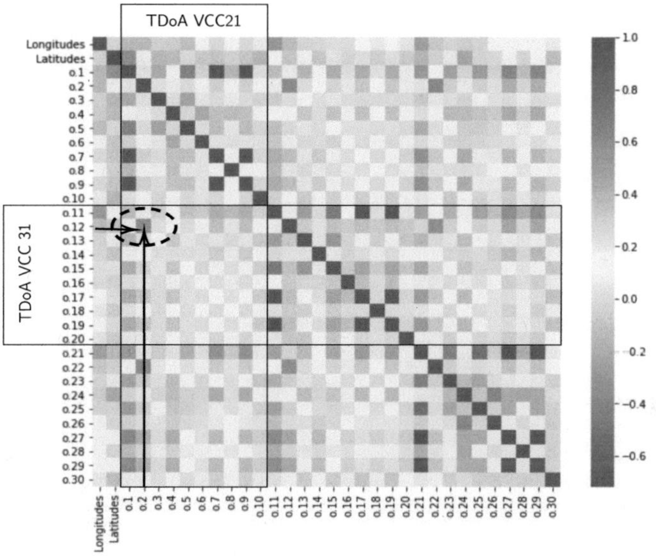


Figure 7.29.: Correlation Analysis of VCC from $\tau_{21}$ and VCC from $\tau_{31}$.

In localization problems, this can give us information on the regions where the localization will have lower performance. In theory, we could select the more useful features depending on the area desired and the scenario characteristics.

This feature selection scheme could be integrated into the localization fingerprints, where we could select features that would contribute most to the quality of the resulting model.

Figure 7.29 shows the correlation between features for the validation scenario. The first 10 features are related to the VCC between TDoA Sensor 1 and 2. As we can observe, features 1 and 2 from $VCC_{21}$ presents a high correlation with features 1 and 2 from $VCC_{31}$.

Based on the investigations carried out in [Biau and Scornet, 2016], we could observe that the high correlation between the features causes lower performance in fingerprint localization.

The Random Forest robustness to deal with the absence of features in the dataset is vital in the localization scheme as we are proposing. The bandwidth constraints and the possibility of multipath specular components closely spaced cause several mismatches during the multipath fingerprints dataset creation.

It was also observed that the power of Random Forest to deal with the absence of data could occur when no multipath component was detected. The treatment of non-linear variations and noise robustness tested in the simulations and suggested by [Louppe, 2014] were also observed.

The use of the linear regression model follows the same steps as the classifier. As discussed in Section 6.2.1, the critical aspect of the proposed model is the production, structuring, and pre-processing of the datasets. Changing the model itself is not very difficult to do.

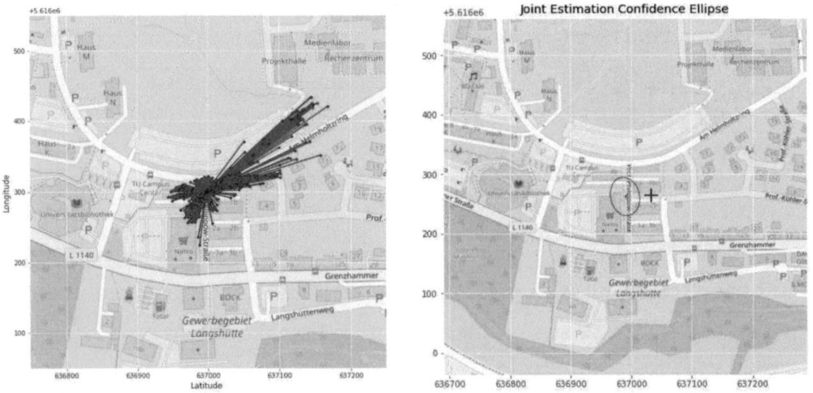


(a) Estimation using Random Forest Classification.
(b) Joint performance using data fusion with ICI.

Figure 7.30.: Enhancement GDOP with RT Fingerprints.

The regression model was configured according to the parameters displayed in Table 7.2 and applied to the datasets. The result of the experimental validation showed superior performance of the Random Forest as a classifier.

The diffuse and highly non-linear nature of the multipath rays makes the binary tree-based classifiers perform better. It is also helpful to remember that we are working with a closed set of possibilities; that is, all points of the simulation grid constitute a *closed convex set*, [Zhang et al., 2016].

The drawback of applying classifiers lies in the size of the models since binary trees with the configured parameters produce models in the order of 30 or 40 GB of memory. In comparison, solutions based on linear regression were in the order of 5 to 8 GB.

Figure 7.30 shows the Random Forest classification and the final result using a data fusion (ICI) approach. Figure 7.31 shows the estimations obtained using the joint approach of TDoA and multipath fingerprints. The final results were fused using the ICI scheme.

Figure 7.31.: Heatmap with Localization Performance of the Multipath Fingerprints using ICI, the yellow parts bad estimation areas.

Figure 7.32 demonstrates the comparative results of the random forest classifier, the linear regression model and a traditional localization system. As we can see the RF outperforms the regression approach, bringing enhancement to the overall localization result.

Another aspect of testing the machine learning model's performance is the accuracy obtained by the fingerprint approach. According to the definitions presented in Appendix B and established by Equation (B.1), the accuracy obtained with the Random Forest model for the scenario described was $86.17\%$, within the average mean square error of 24 meters.

The points where the method's performance has not been satisfactory are the regions where the system itself has also given poor results. Therefore, our strategy of improving existing fingerprint localization systems proved adequate. The use of fingerprints alone can only provide a reasonable estimate in some regions, depending on the quality of the

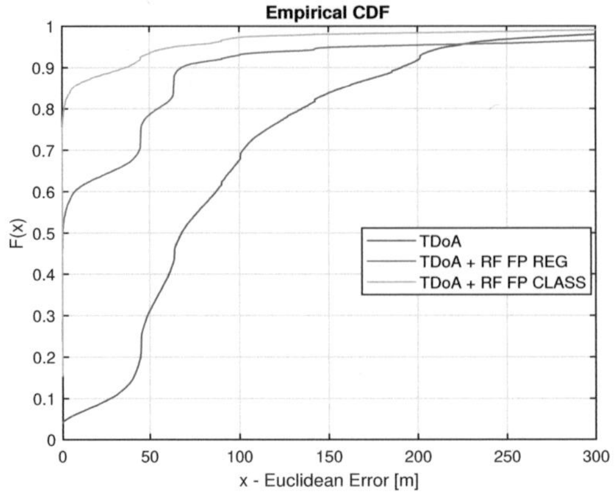


Figure 7.32.: Empirical CDF of the Performance of Classification and Regression.

cross-correlation. Still, in some points, the performance of the method alone would not be appropriate.

The data extraction technique using VCC showed promising results in practice, as cross-correlation is already available in the processing handled by TDOA. It is necessary to remember the inherent limitations of the TDoA systems, regrading the CBB, explored in Section 2.4 because the bandwidth of the system does not allow a high temporal resolution. Therefore, peaks that are very close will be considered a single peak.

In short, the analysis of the results achieved in the validation of the proposed method using GDOP and CDF confirms improvement performance of the localization system. As a general rule, the more real measures are added to the datasets, the better the estimates are. Despite the difficulties in parameter extraction from reference signals, such measures provide an overall improvement of the results.

### 7.3.3. Random Forest Multipath Fingerprints with SIMO FIR

In the city of Uttenreuth took place the final validation scenario. The equipment data-sheets, locations, and the transmitter used are described in the Table 7.7. In this scenario, we use 433 MHz as the operating frequency and a well-defined sequence to assist in the extraction process of the multipath components.

As a rural area, the few buildings and small houses make the specular components

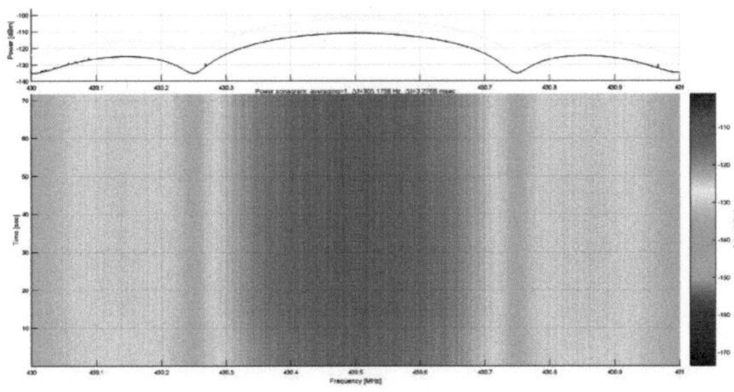

Figure 7.33.: Signal PRBS(15) Received by the TDoA Sensor with 1 MHz of Bandwidth.

more defined. On the other hand, the vegetation and the trees absorbed the signal more than in the previous environments.

Figure 7.33 shows the PRBS(15) received in sensor 1; as we can see, the reception was clean and well-defined during the measurement windows, but the power level is around $-105$ dBm.

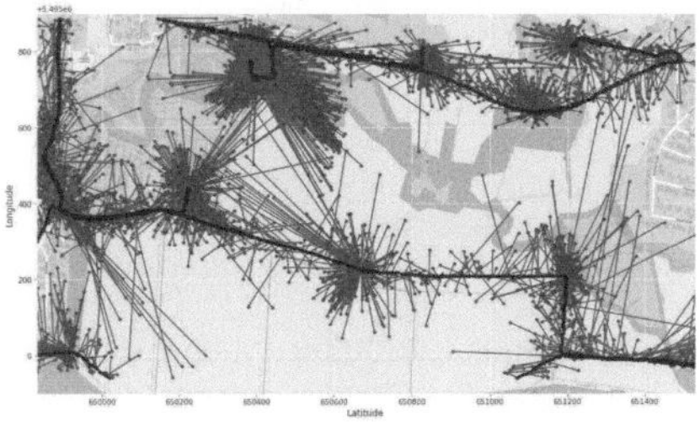

Dataset created using only RT simulations.

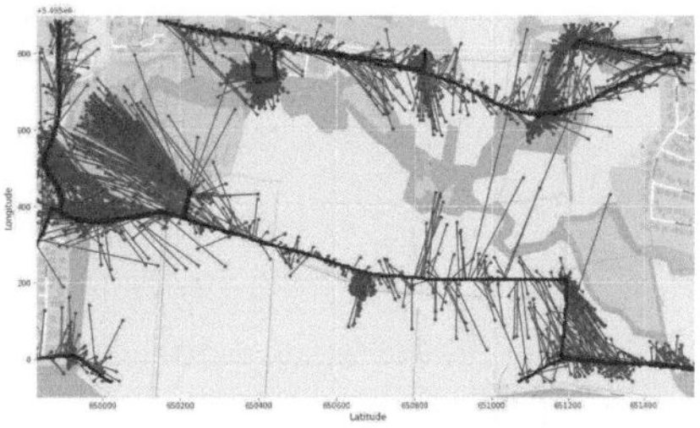

Enhanced Dataset using RT and Reference Emitters with SIMO FIR.

Figure 7.34.: Datasets Enhancement mixing real measurements ans RT simulations.

Once again, the reference emitters were used; at this time, the multipath data were estimated from the received signals using the SIMO FIR technique presented in Section 3.4.

As previously mentioned, the bandwidth limitation caused these estimates not to present a satisfactory result immediately. For this reason, it was necessary to perform several realizations to extract the delay and amplitude components to attach them to the dataset produced by the RT simulation.

The procedure was similar to one used in the ciy of Ilmenau, only the signal processing step for enhancing the dataset was changed, to work with amplitude and delay of the multipath components. The dataset was changed to have 40 features for each point in the grid.

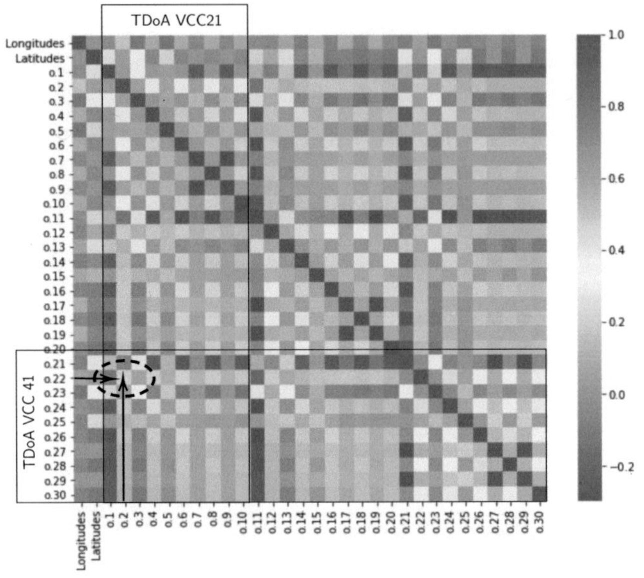


Figure 7.35.: Features correlation analysis.

Figure 7.34 shows the result of the location fingerprint scheme in the Uttenreuth scenario. The blue dots are the ground-truth positions of the reference emitters, and the green dots are the estimates made by the Random Forest. The red lines connecting the two points provide a visual evaluation of the error (Euclidean distance) between the reference point and the estimated one. The left side shows the performance using only the dataset produced with RT. On the right side, we notice estimation improvement when the measurements of the reference emitter using SIMO FIR were added to the dataset.

A feature correlation analysis was also performed to identify the regions where the machine learning prediction could lower performance. Figure 7.35 show the feature correlation diagram.

Figure 7.36 shows the error map produced by the enhanced fingerprints using reference emitter and RT simulation. In this figure, the machine learning model is created, and it is applied in all grid points to estimate the regions of bad performance for localization.

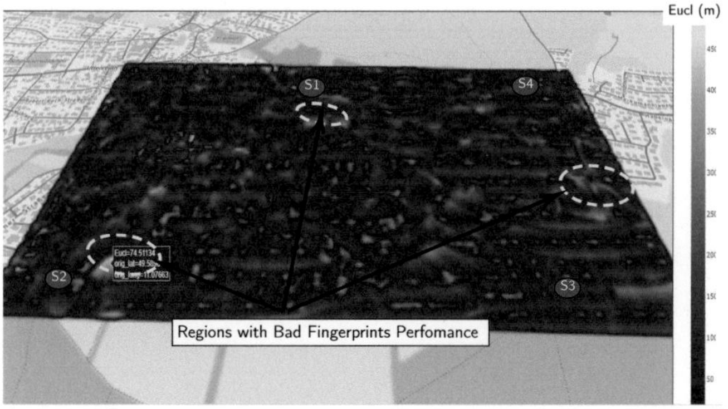


Figure 7.36.: Error map of the RT Multipath Fingerprints Enhanced.

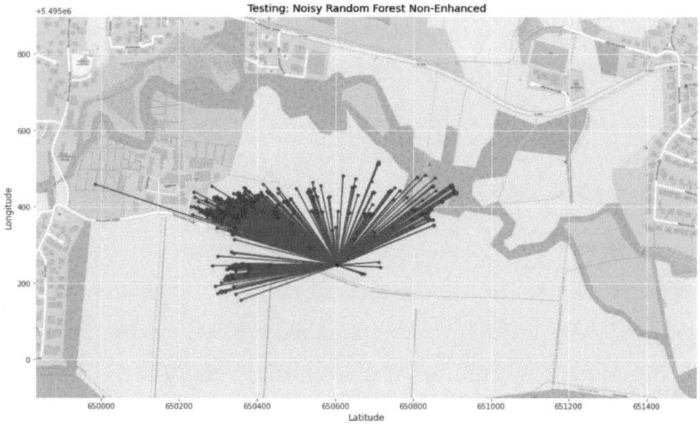


Figure 7.37.: Joint Estimation fusing Multipath Fingerprints and TDoA with ICI.

The regions where the effects of the multipath fingerprint are low are also points where the measurements themselves have bad quality; either the signal was not received, or the NLOS components affected the overall performance. We can see in the map that the Euclidean error is significant. We can compare the positions indicated with the regions of lower performance identified in Figure 7.34.

Figure 7.37 shows the estimations obtained using the joint approach of TDoA and multipath fingerprints. The final results were fused using the ICI scheme.

The final result is presented in the Figure 7.38, which compares the CDF empirical plots of the joint Random Forest using regression and classification with the traditional TDoA systems.

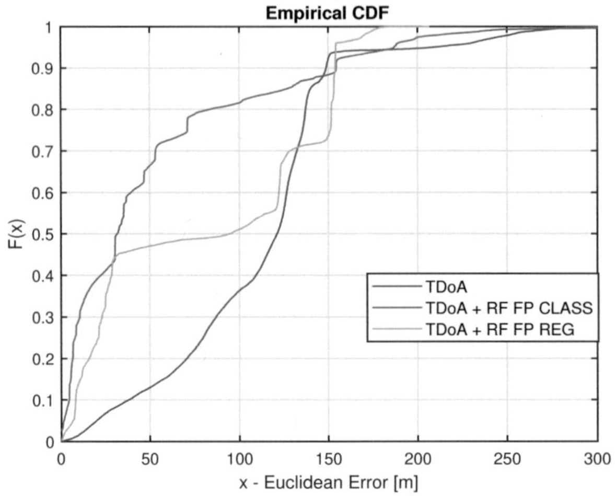


Figure 7.38.: Empirical CDF of the Performance of Classification and Regression.

# 7.4. Analysis of the Results

This section outlines specific aspects of the overall research. Most of the analysis was already presented partially at the end of each simulation. After testing the multipath RT fingerprints using machine learning in a non-collaborative emitter location problem, we can assure that the method works, showing promising results.

The main aspects of the analysis can be grouped as follows:

**Multipath Pre-processing:** the dataset assembly procedure using RT can present a substantial deviation from the real CIR. The low power level signals and the presence of vegetation can make the absorption bigger than the simulated one. As a general rule, the places where the TDoA system could receive the emitter signals were also well represented by the RT simulation. The dataset with zero assignment for the lack of data information in some points did not affect the overall performance. The decision tree formulation of the Random Forest could cope with the mismatching, noise, and lack of multipath information.

**Localization Enhancements:** The multipath fingerprints approach was validated with and without the reference emitters. The hybrid scheme, integrating real measurements in the dataset, showed an overall improvement in the prediction accuracy.

As many real measurements are incorporated in the dataset, as better the prediction model becomes. The localization enhancement was achieved in all the configurations; even with significant fingerprint prediction errors, the overall position estimation was improved.

**Hybrid dataset (RT and reference emitters):** Some real measurements were low quality, meaning that they could not be used either in the TDoA system either in the fingerprint approach. The quality in the cross-correlation is a crucial parameter to be adjusted. If only high-quality measurements are taking into account, the multipath information is lost; if it is too low, it could not be used at all.

**Classification and Regression Performance:** Both Random Forest implementations were tested. Although classification has shown better performance, further investigation is necessary to identify the pros and cons of each solution for the outdoor scenario.

**SIMO FIR or VCC:** The multipath fingerprints were extracted from the measurements using both SIMO FIR and VCC. It seems like the VCC is a more straightforward implementation because the cross-correlation is already in the processing loop of TDoA systems.

The majority of location systems in an outdoor environment use equipment that provides statistical results of the measurements performed. In a real scenario of localization, within a multipath environment, it is not easy to obtain a measure without the influence of NLOS.

Being more pragmatic, to make a connection with the research questions presented in Section 1.3, we can assure from the results showed that the multipath information could be used to improve the performance of outdoor localization systems.

Also, we have found that attaching real measures to the synthetic data produced by the RT simulation can improve the location estimates.

The negative point is the need to extract multipath information from the signals received by the system. Not all recordings were used to improve the dataset, and in addition to the multipath acquisition process, we showed how to extract unfeasible estimates.

As the last remark, we were able to prove that the technique of data fusion, referred to here as ICI, was adequate to gather the fingerprint responses with the actual measurement data.

This dissertation claims that the multipath that affects measurements also affects ray-tracing outcomes and the machine learning model generated; that is, the correlation exists and is difficult to estimate. That is why the strategy of using the ICI proved to be adequate.

## 7.5. Concluding Remarks

This chapter focused on showing the application of the localization method based on a mix of information produced synthetically and by real measures in an outdoor environment.

We studied the method of multipath positioning enhancement using machine learning and multipath fingerprints. Consequently, we evaluated the performance of the proposed scheme in different scenarios. We presented the details of equipment, their configuration, and location conditions used in validation campaigns.

Ilmenau and Uttenreuth were suitable places for the validation of the concept because the propagation has well-defined rays of multi-course, according to the studies presented by [Richter, 2005].

The choice of the features to compose the different datasets followed the guidelines suggested by [Landmann, 2008], showing promising results. We have selected five peaks for both VCC and SIMO FIR in the validation conducted. However, other values could be tested as long as a computational resource is available.

Several aspects of the emitter localization problem were once again revisited during the description of the results. Some signal processing techniques, like ray-tracing and machine learning, were necessary to understand the experiments. The Random Forest classification model showed better results than linear regression. We have used GDOP and CDF to check the performance of the final results, as explained in Appendix A.

In summary, we can state that the localization technique using mixing simulation data with actual measurements and coupling machine learning tools works. The results achieved can be reproduced, provided that they use descriptions of performance-based in mean and variance. As the method presented is based on several realizations, most of the results are better described by statistical moments.

# 8. Conclusions and Outlook

The final chapter reviews the main idea of this thesis and presents future possibilities for using the proposed method in urban and suburban environments. Passive radio frequency localization of emitters plays a vital role in surveillance, intelligence, reconnaissance, and society's safety operations.

The main idea addressed in this study was the use of multipath information to improve outdoor localization systems. Thus, a method was developed that combines ray-tracing propagation data with real-time measurements of reference emitters.

Additionally, machine learning methods were adopted to merge the fingerprint information with the real-time localization systems, resulting in performance improvement.

In previous researches, other approaches were used for acquiring multipath fingerprints, applying different schemes. Nevertheless, we combined the concept of machine learning to extract statistical information from deterministic simulation systems.

A typical machine learning tool works as a *black box* that matches output data with input, thus generating a model that corresponds to data (not used) in the training process. The approach proposed herein requires additional steps when using Random Forest as the machine learning tool and RT models to produce datasets.

The performance of classifiers proved to be a promising technique for outdoor scenarios. Hence, future research should investigate other machine learning algorithms and present comparative studies in such a complex environment.

In the following sections, we present our research contributions, focus on possible future research and other fields of application, and final remarks.

## 8.1. Key Points and Contributions

Locating mobile stations in a wireless environment with increasing accuracy is still one of the main challenges of today's positioning systems. Among the available techniques,

Radio Frequency Fingerprinting (RF fingerprinting) shows promising results due to its excellent ability to estimate the location of mobile stations.

In the first chapter, the general idea of multipath fingerprints was presented by using a machine learning tool, as described in Figure 1.4. The motivation, scope, definitions, and state of the art in the proposed research were presented in detail. In Chapter 2, we reported not only the fundamentals of TDoA radio localization but also explained the practical limitation effects of bandwidth, SNR, synchronism requirements, and integration time on system precision.

In Chapter 2, we revised the mathematical formulation, localization algorithm, and practical constraints of TDoA localization systems. Moreover, multipath effects in the system's performance were examined with an emphasis on the impact of bandwidth, $SNR$, synchronism requirements, and integration time on system precision.

In Chapter 3, we showed a signal processing scheme for blind channel impulse estimations techniques. These techniques are the core of the hybrid method proposed tool for connecting RT with real measurements. We investigated blind SIMO FIR and VCC. The critical aspects of the CIR in the localization framework were explored in detail. Furthermore, we evaluated the effects of reduced bandwidth in the CIR estimations and showed the improvement achieved with long recordings to cope with lower time resolution.

RT simulation tools are designed for analyzing communication system coverage. Therefore, it was necessary to modify it for the multipath fingerprints production. Chapter 4 explained the adopted procedures and the technical aspects behind the computational tool. This perception was essential to know the grid method to divide the scenario and produce the multipath signatures. We also explained the visibility tree, a trade-off of the collecting region size, and its impact on the fingerprints.

Chapter 5 illustrated the fundamental aspect of machine learning in general and the adaptation required to deal with electronic localization problems. Additionally, we described the basic differences between regression and classification set up to produce the estimation models.

The Random Forest method was explained both by decision trees and the mathematical basis of the linear regression and classification models applied to the multipath components. This chapter concluded the first part of the thesis, integrating the main aspects of outdoor localization, channel estimation techniques, and machine learning tool applicable to the problem.

Starting the second part of this work, Chapter 6 described the proposed method, presenting the scheme as a composition of the issues addressed in Part 1 of the thesis. Besides, we discussed data augmentation techniques using the unification of measurements as an expanded data set within model training.

Part 2 of the research concludes with the experimental validations presented in Chapter 7. We also put various scenarios to validate the strategy of a database of TDoA and Random Forest used as classifiers and as linear regression.

Several aspects of research contributions can be found throughout the thesis. Practical features of localization systems were explored in Chapter 2. As seen in Chapter 3, we have proposed two new schemes to extract the multipath patterns from the channel estimation. We also introduced a modified approach for using RT with Random Forest, fusing the fingerprints predictions with real measurements. In summary, we have proposed to employ tools, usually applied in other areas, to build the multipath exploitation scheme.

More specifically, the extraction of multipath patterns contained in the cross-correlation function, the use of VCC in this application can be considered a significant contribution. Moreover, if implemented in practice, such a procedure can use available information, adding one extra step, improving localization systems. Another contribution to the area is the scheme of using RT for a synthetic dataset of the VCC fingerprints.

In summary, we conclude that the current research is helpful in the area of electronic localization in outdoor environments. Despite the limitations of CIR estimation, the method presented has shown promising results, combining ray-tracing with machine learning in a localization problem.

## 8.2. Future Works

Based on our research and scientific contributions, future research work can lead to many possible directions. We envision three main opportunities in the field of machine learning and the other two possibilities related to outdoor radio-frequency measurements. The list of the research steps are as follows:

**Deep Learning Models (DNN).** Our research focuses on random forest models. However, machine learning algorithms require structured data. In contrast, deep learning networks can operate in unstructured data, relying on layers of artificial neural networks. Hence, the performance of other methods will be tested and evaluated compared to the current Random Forest technique.

**Use of Cross-Correlation Shape instead of Peaks.** We incorporated the Random Forest in the classification of data and linear regression due to the natural connection between our datasets and the multipath information. In this scheme, we have used five peaks of the VCC or five rays of the CBB in each sensor. With this procedure, the dimension of incident rays will reduce in size because it is possible to obtain much more suppressed information about multipath patterns and reflections compositions in the shape of the VCC. Moreover, in future work, one should concentrate on producing a dataset with the VCC function extracting more features in the positioning fingerprints.

**Use of other ML Models: KNN and CNN.** We have used the Random Forest model in classification and linear regression analysis due to the natural connection between the information of rays and the decision trees. However, each method needs to be analyzed to open the "black box" and understand the mathematical foundation. Hence, it is required to properly tune the parameters when applying the idea of the radio-frequency fingerprints.

**Use of AoA and FDoA in the Datasets.** We have only used the amplitude and delay of the incident rays $(\alpha_i, \tau_i)$. However, both the localization system and the RT simulations provide angular information as well, and it is possible to use the full polarimetric CIR, described by [Richter, 2005] in the fingerprints formulation.

**Use of Complex Ambiguity Function (CAF) in the Datasets.** We have used the TDoA to give us the position estimations. Still, the same setup would enable us to extract the time-delay features and the effects of Doppler delay. Hence, both pieces of information will give us a CAF fingerprint of the whole position emitter vector, including the speed.

In summary, possible developments of future research shall address problems in the field of machine learning, simulation tools, or measurements of the related parameters. We discussed a few future directions here. However, the application of these tools is not limited to this particular method. Outdoor localization is a vast field, and comparative studies between the different techniques will always produce exceptional results.

## 8.3. Other Fields of Application

Throughout the research, we have focused on a non-collaborative emitter localization in an outdoor situation. However, this application scenario can be expanded to various communication systems. For example, the topic of RF signal fingerprinting can be applied to problems of the Internet of Things (IoT), to automobile self-guidance systems, or even to problems involving passive object detection, also called multipath exploitation radar [Leigsnering et al., 2014] and [Setlur et al., 2011].

In the field of Internet of the Things (IoT), it is usual to have a trade-off between the places available for the installation of sensors and the area of interest. The use of multipath location techniques can bring benefits because it would be possible to reach an adequate level of information accuracy, even using a few sensors. Thus, in pier containers, or large warehouses, the multipath conditions do not change significantly over time. Therefore, it is a suitable place for the application of fingerprints based on machine learning.

In the segment of car automatic-guidance systems, there is potential for applying our research scheme. In this field, positioning and obstacle detection is a necessary factor. Hence, using multipath information can assist other systems in improving the performance of already installed equipment.

Moreover, machine learning techniques integrated with the extracted fingerprint data of both TDoA and Frequency-Difference Of Arrival (FDoA) can increase situational awareness of moving vehicles.

The Passive Coherent Localization framework, introduced by [Thoma et al., 2019], can also take benefit of the RT machine learning fingerprints idea. In the case of new services in 5G systems, this method can enhance the passive radar performance. In this case, the position processing could be added to the proposed software-defined radio architecture. Therefore, the application of this research will extend in other fields, including applications in industry, transport, logistics, and safety and security situations.

## 8.4. Final Remarks

The research area of emitter localization using a network of distributed sensors is considered an essential part of various systems in outdoor and indoor environments. Moreover, there are many open research questions, which need to be studied and solved.

The strategies that assume LOS condition between the sensors and the emitter are not feasible in outdoor urban scenarios. The multipath effects are inherent to the outdoor environment in the real world and are very challenging to eliminate.

Some filtering techniques may be employed to avoid NLOS measurements. But in many situations, this method is not appropriate due to the inherent channel components. Only in a few applications, the LOS condition is accomplished. Therefore, it is necessary to understand the multipath effects, and to some extent, we may enhance the channel characteristics. Among diverse possible strategies, this thesis was devoted to exploring the potential of RT tools in conjunction with machine learning to improve the overall performance of the system.

Our scheme showed that the multipath patterns information provides an improvement in the location estimation. The results obtained indicate a promising use of machine learning techniques in conjunction with RT. However, more testing and validation studies are still needed.

Finally, this study contributes to many research findings by developing a new approach to improve localization problems. Other potential areas can also take benefit from the deployment of different systems in an outdoor environment.

# Appendices

# Appendix A

# Performance Criteria in Localization Systems

This appendix explains the derivation of two classes of performance criteria for localization systems. The first class comprises the Geometric Dilution of Precision (GDOP) or Confidence Ellipsis and the Circular Error Probability (CEP); the second class is the Cramér-Rao Lower Bound (CRLB), considering the noise as Gaussian.

The mean and the variance of the errors in the position estimation describes the achievement of Time-Difference Of Arrival (TDoA) systems and are relevant parameters to compare different implementations using statistical bounds, such as the Cramer-Rao lower bound (CRLB). We used the error here as a benchmark parameter to evaluate the improvement of using several different localization techniques.

First, we assume that the exact emitter position is available, either from a reference emitter or a known target situation. This could be a valid approach, but we deal with a non-collaborative emitter in an unknown position regarding the scope of this research.

The actual emitter position is given by a vector $\boldsymbol{x}$, while the estimated one is assumed to be a vector $\hat{\boldsymbol{x}}$. We define the mean position error as a vector that describes the the bias $\mathbf{b}$, between the estimated radio-frequency target position and the actual one, as:

$$\mathbf{b} = E\left\{(\hat{\boldsymbol{x}} - \boldsymbol{x})\right\}. \tag{A.1}$$

The covariance matrix $\mathbf{C}_{\hat{x}}$ of errors between the estimated and exact positions is:

$$\mathbf{C}_{\hat{x}} = E\left\{(\hat{\boldsymbol{x}} - \boldsymbol{x} - \mathbf{b})(\hat{\boldsymbol{x}} - \boldsymbol{x} - \mathbf{b})^T\right\}. \tag{A.2}$$

## A.1 Cramer Rao Lower Bound

In estimation theory, it is necessary to define a parameter to compare different methods; the CRLB represents the lower bound on variance achieved by any unbiased estimators. It is also possible to use the Empirical Cumulative Distribution Function (CDF) of the positioning error.

[Torrieri, 1984] shows an in-depth calculation of CRLB based on the TDoA measurements matrix. Hence, we compiled here the essential aspects of the outdoor positioning, assuming a zero-mean Gaussian noise. The first step is to obtain the inverse of the Fisher Information Matrix (FIM), where the diagonal elements are the minimum variance achieved by the estimator.

Defining $\mathbf{I}(\boldsymbol{x})$ as the FIM, it is possible to compute its inverse, when the error measurements are zero-mean Gaussian distributed, as in [Chan and Ho, 1994]. The process to formulate the CRLB includes:

1. Obtain the second-order derivatives of the logarithm of the measurement Probability Density Function (PDF) with respect to $\boldsymbol{x}$, that is:

$$\partial^2 \ln p(\boldsymbol{r})/(\partial \boldsymbol{x} \partial \boldsymbol{x}^T). \tag{A.3}$$

2. Obtain the expected value of $\partial^2 \ln(\boldsymbol{r})/(\partial \boldsymbol{x} \partial \boldsymbol{x}^T)$ so,

$$\mathbf{I}(\boldsymbol{x}) = E\{\partial^2 \ln(\boldsymbol{r})/(\partial \boldsymbol{x} \partial \boldsymbol{x}^T)\}. \tag{A.4}$$

3. The CRLB for 3D coordinate is: $[\mathbf{I}^{-1}(\boldsymbol{x})]_{1,1}, [\mathbf{I}^{-1}(\boldsymbol{x})]_{2,2}, [\mathbf{I}^{-1}(\boldsymbol{x})]_{3,3}$

4. The matrix $\mathbf{I}$ is calculated by:

$$\mathbf{I}(\boldsymbol{x}) = [\frac{\partial \mathbf{f}(\boldsymbol{x})}{\partial \boldsymbol{x}}]^T \mathbf{C}^{-1} [\frac{\partial \mathbf{f}(\boldsymbol{x})}{\partial \boldsymbol{x}}], \tag{A.5}$$

The covariance matrix for the noise $\boldsymbol{n}$ is $\mathbf{C}$, as seen in Equation (A.7), and using the definition of $\mathbf{f}_{TDOA}(\boldsymbol{x})$ introduced by [Zekavat and Buehrer, 2011]:

$$\left[\frac{\partial \mathbf{f}_{TDOA}(\boldsymbol{x})}{\partial(\boldsymbol{x})}\right] = \begin{bmatrix} \left[\frac{x-x_2}{d_2} - \frac{x-x_1}{d_1}\right] & \left[\frac{y-y_2}{d_2} - \frac{y-y_1}{d_1}\right] & \left[\frac{z-z_2}{d_2} - \frac{z-z_1}{d_1}\right] \\ \left[\frac{x-x_3}{d_3} - \frac{x-x_1}{d_1}\right] & \left[\frac{y-y_3}{d_3} - \frac{y-y_1}{d_1}\right] & \left[\frac{z-z_3}{d_3} - \frac{z-z_1}{d_1}\right] \\ \vdots & \vdots & \vdots \\ \left[\frac{x-x_L}{d_L} - \frac{x-x_1}{d_1}\right] & \left[\frac{y-y_L}{d_L} - \frac{y-y_1}{d_1}\right] & \left[\frac{z-z_L}{d_L} - \frac{z-z_1}{d_1}\right] \end{bmatrix}, \tag{A.6}$$

and,

$$CRLB_{TDoA}(\boldsymbol{x}) = [I^{-1}_{TDoA}(\boldsymbol{x})]_{1,1} + [I^{-1}_{TDoA}(\boldsymbol{x})]_{2,2} + [I^{-1}_{TDoA}(\boldsymbol{x})]_{3,3}. \tag{A.7}$$

The CRLB is determined according to Equation (A.7). The computation requires the error covariance matrix, which is generally unavailable for TDOA algorithms, particularly for iterative methods such as the Least Square (LS) and the gradient descent.

## A.2 Confidence Error Ellipse

The confidence error ellipse is a standard tool to evaluate a statistical process, also useful to measure position accuracy. In two dimensional case, the error ellipse can evaluate the distribution of estimates at the bias point, with the following structure:

$$\left(\frac{x - b_x}{\sigma_x}\right)^2 + \left(\frac{y - b_y}{\sigma_y}\right)^2 = \gamma, \tag{A.8}$$

$b_x$ and $b_y$ are the $x$ and $y$ components of the bias vector $\boldsymbol{b}$, $\sigma_x, \sigma_y$ are the $x$ and $y$ variance terms from $\mathbf{C}_{\hat{x}}$, respectively, and $\gamma$ is a scale parameter used to determine the area of the ellipse.

Parameter $\gamma$ controls the percentage of estimates that are within the ellipse, defining its confidence interval:

$$P\{\hat{z} < \gamma\} = 1 - e^{-\gamma/2}, \tag{A.9}$$

$$\gamma = -2\ln(1 - P), \tag{A.10}$$

where $\sigma_s$ is the smallest of the two error terms and $\sigma_l$ is the larger one. The first approximation is valid when the error ellipse has low eccentricity (the two error terms are approximately equal), and the latter is used when one error term dominates.

Following the derivation presented in [Cline, 1978] and [O'Donoughue, 2019], we can use the dummy variables $\tilde{x}, \tilde{y}$, replacing the $x, y$. The ellipsis has the following structure:

$$\begin{aligned} \alpha &= \tan^{-1}(v_y/v_x), \\ x &= \cos(\alpha)\tilde{x_1} - \sin(\alpha)\tilde{x_2}, \\ y &= \sin(\alpha)\tilde{x_1} + \cos(\alpha)\tilde{x_2}. \end{aligned} \tag{A.11}$$

The covariance matrix of the TDoA is not diagonal, because there is a correlation of the measurements with the base-sensor. Consequently, Equation (A.8) uses the square root of the eigenvalues of the covariance matrix $\sqrt{\lambda_1}$ and $\sqrt{\lambda_2}$, instead of the standard deviation $\sigma_x$ and $\sigma_y$. These eigenvalues establish the minor and major axis of the confidence ellipse.

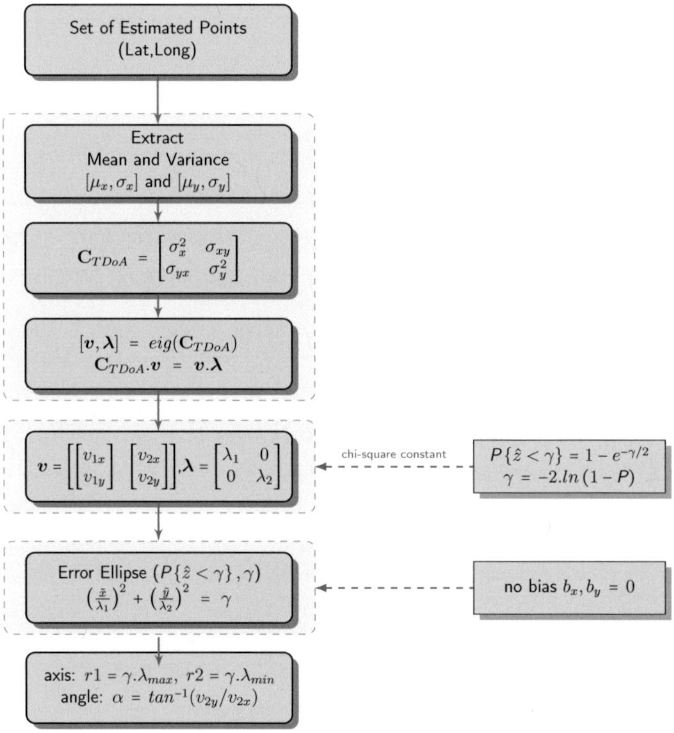


Figure A.1: Geometric Dilution of Precision (GDOP) Formulation.

The eigenvector $\boldsymbol{v}$ with the first eigenvalues $\lambda_x, \lambda_y$ are used to give the $x, y$ components in the ellipsis equation. The parameter $\gamma$ defines the confidence interval and can be taken from tables of $chi - squared$ distribution; some of these values are given in Table A.1.

Table A.1: Error Ellipse Scale Factor and Confidence Interval

| Confidence Interval | Scale Factor $\gamma$ |
|---|---|
| 39.35% | 1.000 |
| 50.00% | 1.386 |
| 63.20% | 2.000 |
| 75.00% | 2.773 |
| 90.00% | 4.601 |
| 95.00% | 5.991 |
| 99.00% | 9.210 |

Figure A.1 shows the confidence ellipsis calculation to perform the Geometric Distribution of Precision GDOP, including the minor axis, principal axis, and inclination angle.

## A.3 Circular Error Probability (CEP)

The CEP defines a circular region, with the emitter in the center, that gives the percentage of estimates. It usually is scalar bidimensional. It is also possible to include all three dimension error terms, resulting in a spherical error probable (SEP).

The value CEP is given by the radius of a circle that encloses errors with a given probability. Assuming that the total position error is $\tilde{z}$

$$\tilde{z} = \sqrt{(\hat{x})^2 + (\hat{y})^2}. \tag{A.12}$$

The CEP is obtained by the following expression:

$$\int_0^{CEP_\gamma} f_z(\tilde{z}) = \frac{\gamma}{100}. \tag{A.13}$$

The Equation (A.13) do not have a closed-form solution, but it is possible to used an approximation [O'Donoughue, 2019].Assuming that $\sigma_s, \sigma_l$ as the shortest and the largest error in the estimation , the usual approximation to obtain the $\text{CEP}_{50}$ can be given by Equation (A.14).

$$\text{CEP}_{50} = \begin{cases} 0.59(\sigma_s + \sigma_l)) & \frac{\sigma_s}{\sigma_l} \geqslant 0.5 \\ \sigma_l(0.67 + 0.8\frac{\sigma_s^2}{\sigma_l^2}) & \frac{\sigma_s}{\sigma_l} < 0.5 \end{cases} \tag{A.14}$$

In the case of 50% [Ignagni, 2010], the notation used is $\text{CEP}_{50}$, but can also make an approximation for other values:

$$\text{SEP}_{50} \approx 0.51\left(\sigma_x + \sigma_y + \sigma_z\right), \tag{A.15}$$

It was also possible to analyze the localization system performance in 3D, using spherical error probable (SEP), an extension of the CEP formulation.

In this case, it is necessary to use the approximation based on the relative value of the standard deviation of error in the three dimensions, as explained by In [O'Donoughue, 2019]. The approximations for SEP in 50 and 90 percent, respectively, are:

$$\text{SEP}_{90} \approx 0.833\left(\sigma_x + \sigma_y + \sigma_z\right). \tag{A.16}$$

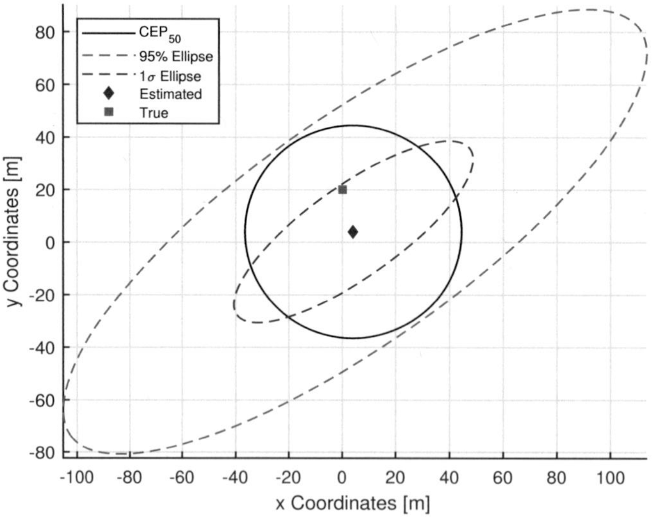


Figure A.2: $\text{GDOP}_{95}$ and $\text{CEP}_{50}$ in TDoA systems

## A.4 Empirical Cumulative Distribution

In statistics, the CDF is the distribution function associated with a defined empirical measure of an experiment that can also be used to compare different localization algorithms. The distribution function of the errors provides a tool to compare which implementation gives a better result, [Ignagni, 2010]. In short, the localization process is typically statistical.

Sometimes, the distribution of observations for an error in the estimation does not fit any known probability distribution. The bias corrupts the results in multipath errors, and it is difficult to apply a "curve fitting" in real-world datasets.

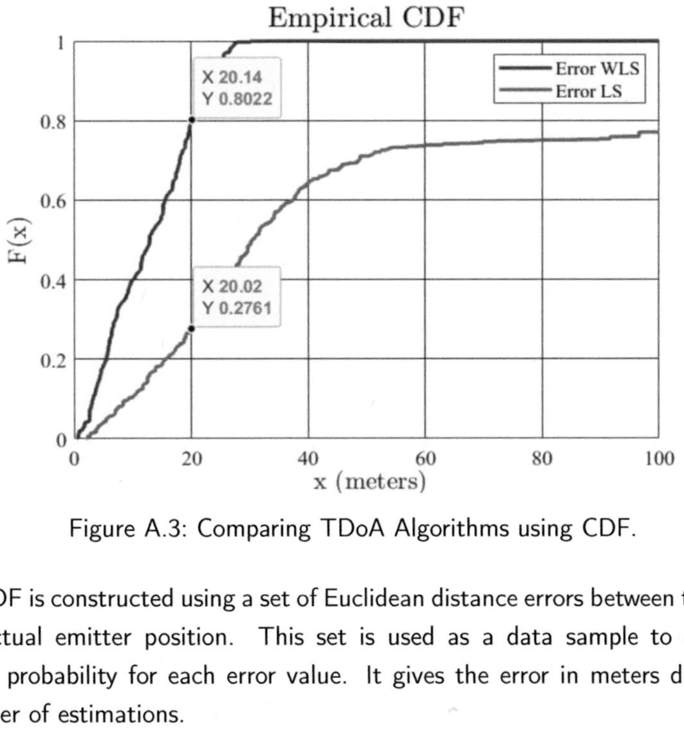


Figure A.3: Comparing TDoA Algorithms using CDF.

The CDF is constructed using a set of Euclidean distance errors between the estimated and the actual emitter position. This set is used as a data sample to calculate the cumulative probability for each error value. It gives the error in meters divided by the total number of estimations.

The Figure A.3 shows the CDF of a localization system using Weighted Least Square (WLS) (blue line) and LS (red line). If we define an error of 20 meters, the LS approach has only 27,6%, and WLS has 80% of the estimation are lower than this value. Both methods used a Signal to Noise Ratio (SNR) of 10 dB, using the formulation presented in Equation (2.17) and Equation (2.32).

The Figure A.2 shows the result for the WLS method, where it is shown the comparison between the $CEP_{50}$ and a GDOP confidence Ellipsis with 95%. It can be mentioned that these tools give a visual performance comparison of different localization approaches.

In the simulation exhibited in Figure A.2, we used a "bias" to force the multipath effect $\boldsymbol{b}$ of 100 meters in the time-difference $\tau_{21}$ and a zero-mean Gaussian noise. The results show that Linear Least Square (LLS) with 10 dB presents better performance than WLS.

# Appendix B

# Machine Learning Terminology

In this section, we discuss the machine learning taxonomy used in the proposed method. Although machine learning has well-established terminology in the technical literature [Raschka, 2015], we evoked the usual terms, associating them to our ray-tracing fingerprints framework.

**Dataset:** The term dataset represents a collection of available data to train and evaluate the network. In the proposed work, the dataset is collected by Ray Tracing (RT) simulations and few measurement points, which are assumed to be calibrators or reference emitters. We split the outdoor environment into a grid of $N$ points, where each coordinate and multipath pattern features represent one element. Generally, the dataset is split into training, validation, and test sets. An instance of a dataset represents a row in the dataset table, also known as observation, whereas the feature is a column in the dataset table.

**Features and Labels:** A dataset is usually presented in the form of a table that contains features and the respective targets. In this work, the components comprise the multipath Channel Impulse Response (CIR), a pair of amplitude-delay $\alpha, \tau$ or the peaks in the Volume Cross-Correlation (VCC) function. The labels represent the elements that are to be predicted by the machine learning algorithm. In our case, the targets constitute the possible positions of the emitter.

**Model:** The model is a weight matrix that can map the input features to the target labels. In the training step, we optimize this weight matrix to minimize the error between the predicted outcome and the actual outcome.

**Training Set:** The training dataset represents a subset of the dataset. This subset is used to train or tune the machine learning model parameters or weight matrix.

**Validation Set:** The validation dataset is part of the dataset used to validate the trained model. Please note that the validation set comprises different samples from the training set, i.e., none of the training samples in the validation set is a part of the training dataset.

**Test Set:** The test set represents the subset of the dataset used to test the trained model. In other words, the test set provides the means to evaluate the trained weight matrix.

**Algorithm:** An algorithm collects sequential rules that enable the machine learning technique to perform a defined task by mapping inputs to an output. Machine Learning uses a set of different methods and allows computers to learn from data and improve predictions.

**Exploration and Exploitation:** *exploration* and *exploitation* are two terms that are often used to describe the ability of the machine learning model to estimate the correct position of the source in localization problems. These two terms have slightly different meanings [Osugi et al., 2005]. The exploration refers to the capacity of an algorithm to reach other promising regions in the search space. Conversely, exploitation is related to the algorithm's performance for searching optimal solutions within the given areas. In a broader sense, exploitation is an intense operation to improve the first approach, whereas exploration is the diversification operation that tries to find the optimal solution.

**Confusion Matrix:** The confusion matrix is used to evaluate the classifier's performance or the trained model. An example of a confusion matrix is presented in Table B.1. In this work, we use exploitation for probing a limited search space region for improving promising solutions that are already pre-selected. Let us consider that we have $N$ classes. Then, a confusion matrix is $N \times N$ matrix that maps the actual outputs to the predicted results.

In Table B.1, the $1^{st}$ quadrant (a) is the true positive, i.e., where a model predicts the position correctly. Quadrant (b) is the false positive, i.e., the observation is negative, whereas the model predicted positive. Quadrant (c) is the false negative, i.e., the observation is positive; however, the model predicted negative. Quadrant (d) predicts a true negative, i.e., the observation is negative, and the prediction is also negative.

**Accuracy:** The term accuracy is defined by the ratio of the correct predicted classes to

Table B.1: Confusion Matrix

| | | Predicted | |
|---|---|---|---|
| | | Good | Bad |
| Actual | Good | True Positive (a) | False Negative (c) |
| | Bad | False Positive (b) | True Negative (d) |

the total number of predictions. The accuracy of any classifier is determined by the following relation:

$$\frac{a+c}{a+b+c+d}. \tag{B.1}$$

There are several different implementations of machine learning algorithms. The terms mentioned above are used in the machine learning formulation. It is noticeable that the combination of different machine learning algorithms makes the predictions more precise.

# Appendix C

# Fusion Estimation in Localization Problems

This section describes the fusion mechanism that integrates the ray-tracing multipath fingerprint estimations with an actual TDoA system. We introduce a concept for position estimation data fusion, which is the strategy to enhance the localization system based on machine learning results with the RT fingerprints.

In the localization problem, the position estimation from different methods can be modeled as a data fusion problem [Koch, 2016], using the statistical descriptions of each approach. The final step in the framework proposed is to show how to perform a consistent position estimation fusion. Therefore, this appendix explains the multipath fingerprints setup used in this work.

The data fusion can use a centralized or a decentralized approach; consequently, the multipath fingerprints dataset calculated in the master TDoA sensors drives us to adopt the centralized one.

A classical fusion problem is how to estimate the cross-correlation information among the different data. In our case, there is some grade of inter-dependency between the multipath effects in TDoA measurements and RT fingerprints.

There are several ways to fuse two different estimations, some more limiting [Julier and Uhlmann, 2017] than others because the central question is how to deal with unknown cross-correlations [Hlinka et al., 2014].

This unknown correlation may lead to an estimate that is optimistic in terms of error covariance matrices. The joint position information to be fused is often given in local state estimates and uses the corresponding estimated error covariance matrices.

As described in [Noack et al., 2017a], a position data fusion is a challenging task due to the difficulty to maintain and track the information required to perform an optimal fusion. Either one takes into account too much information or discards some useful one.

One possibility to deal with unknown cross-covariance data fusion is the covariance intersection (CI) algorithm, which conservatively fuses local estimates. In the multipath fingerprints data fusion, we have adopted the Inverse Covariance Intersection (ICI) to perform the joint estimation.

The ICI presented by [Noack et al., 2017a] offers a less conservative fusion rule. It uses a minimization process to find the cross-correlation between the estimations [Noack et al., 2017b].

The fusion rule between the TDoA system and RT Fingerprints measurements can adopt the ICI fusion rule. Let the TDoA estimation be represented by the mean and variance $(\hat{\mu}_{TDoA}, \hat{\mathbf{C}}_{TDoA})$, and the RT Multipath Fingerprints estimation be represented by $(\hat{\mu}^{FP}, C^{FP})$ the measurements, the ICI fused estimation is given by [Reinhardt et al., 2015] and [Noack et al., 2017a] as follows:

$$\hat{\mu}_{ICI} = \mathbf{K}_{ICI}\hat{\mu}_{TDoA} + \mathbf{L}_{ICI}\hat{\mu}_{FP}, \tag{C.1}$$

$$\mathbf{C}_{ICI}^{-1} = \mathbf{C}_{TDoA}^{-1} + \mathbf{C}_{FP}^{-1} + (-\omega\mathbf{C}_{TDoA} + (1-\omega)\mathbf{C}_{FP}))^{-1} \tag{C.2}$$

$$\mathbf{K}_{ICI} = \mathbf{C}_{ICI}\left(\mathbf{C}_{TDoA}^{-1} - \omega\left(\omega\mathbf{C}_{TDoA} + (1-\omega)\,\mathbf{C}_{FP}\right)^{-1}\right), \tag{C.3}$$

$$\mathbf{L}_{ICI} = \mathbf{C}_{ICI}\left(\mathbf{C}_{FP}^{-1} - (1-\omega)\left(\omega\mathbf{C}_{TDoA} + (1-\omega)\,\mathbf{C}_{FP}\right)^{-1}\right). \tag{C.4}$$

Figure C.1 shows a scheme for the ICI, indicating the fusion rules applied with the unknown cross-correlation. The blue sphere illustrates the conservative covariance intersection data fusion.

As derived by [Noack et al., 2017a], it is possible to achieve consistent results with the ICI approach without the computation of the cross-covariances.

In short, the localization enhancement proposed in our research uses an ICI fusion architecture to deal with the uncertain quantities with unknown cross-correlations between

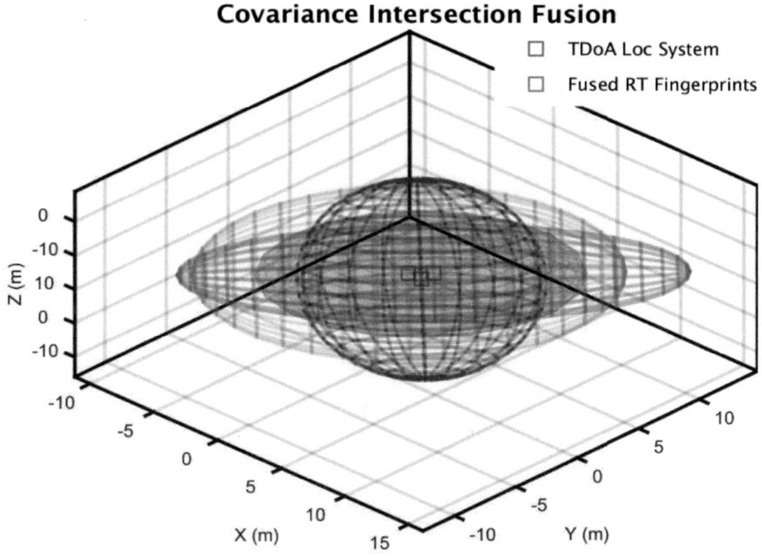


Figure C.1: ICI Fusion.

the ray-tracing fingerprints and the natural TDoA measurements.

We have addressed the enhancement operation as a joint fusion approach for localization in a hybrid system. Once the enhancement is the core of the proposed method, the fusion approach is a crucial aspect of the final solution.

# Bibliography

## Own Publications

[CA15104 and de Sousa, 2018] CA15104, C. A. and de Sousa, M. N. (2018). Whitepaper on new localization methods for 5g wireless systems and the internet-of-things. Technical Report COST Action CA15104, IRACOM.

[de Sousa et al., 2018] de Sousa, M. N., Corrêa, E. F., and Thomä, R. S. (2018). *Time Difference of Arrival Enhancement with Ray Tracing Simulation*, pages 1113–1123. Springer International Publishing, Cham.

[de Sousa et al., 2021] de Sousa, M. N., Sant´Ana, R., Fernandes, R. P., Duarte, J. C., and Thomä, R. R. S. (2021). Improving the performance of a radio-frequency localization system in adverse outdoor applications. *EURASIP Journal on Wireless Communications and Networking*, 2021(1):1–26.

[de Sousa and Thomä, ] de Sousa, M. N. and Thomä, R. S. Mobile station localization emitter in urban nlos using multipath ray tracing fingerprints and machine learning. In *8th International Conference on Localization and GNSS, ICL-GNSS 2018, Guimarães, Portugal, June 26-28, 2018.*

[de Sousa and Thomä, 2018a] de Sousa, M. N. and Thomä, R. S. (2018a). Enhanced vehicle localization in urban NLOS using multipath ray tracing fingerprints. In *88th IEEE Vehicular Technology Conference, VTC Fall 2018, Chicago, IL, USA, August 27-30, 2018*, pages 1–5. IEEE.

[de Sousa and Thomä, 2018b] de Sousa, M. N. and Thomä, R. S. (2018b). Enhancement of localization systems in NLOS urban scenario with multipath ray tracing fingerprints and machine learning. *Sensors*, 18(11):4073.

[de Sousa and Thomä, 2018c] de Sousa, M. N. and Thomä, R. S. (2018c). Localization of UAV in urban scenario using multipath exploiting tdoa fingerprints. In *29th IEEE Annual International Symposium on Personal, Indoor and Mobile Radio Communications, PIMRC 2018, Bologna, Italy, September 9-12, 2018*, pages 1394–1399. IEEE.

[de Sousa and Thomä, 2018d] de Sousa, M. N. and Thomä, R. S. (2018d). Single sensor RF emitter location using ray tracing multipath exploitation. In *2018 15th International Symposium on Wireless Communication Systems (ISWCS)*, pages 1–6, Lisbon,

Portugal.

[de Sousa and Thomä, 2018e] de Sousa, M. N. and Thomä, R. S. (2018e). Time difference of arrival enhancement with ray tracing simulation. In *WSA 2018; 22nd International ITG Workshop on Smart Antennas*, pages 1–5, Bochum, Germany.

[de Sousa and Thomä, 2019a] de Sousa, M. N. and Thomä, R. S. (2019a). Applying random forest and multipath fingerprints to enhance TDOA localization systems. *IEEE Antennas and Wireless Propagation Letters*, PP:1–1.

[de Sousa and Thomä, 2019b] de Sousa, M. N. and Thomä, R. S. (2019b). Enhanced localization systems with multipath fingerprints and machine learning. In *30th IEEE Annual International Symposium on Personal, Indoor and Mobile Radio Communications, PIMRC 2019, Istanbul, Turkey, September 8-11, 2019*, pages 1–6. IEEE.

[de Sousa and Thomä, 2020] de Sousa, M. N. and Thomä, R. S. (2020). *Machine Learning and Multipath Fingerprints for Emitter Localization in Urban Scenario*, pages 217–230. Springer Singapore, Singapore.

# References by Other Authors

[Aatique, 1997] Aatique, M. (1997). *Evaluation of TDOA techniques for position location in CDMA systems*. PhD thesis, Virginia Tech.

[Abdulkabir, 2019] Abdulkabir, B. (2019). *Radio Frequency Toolbox For Drone Detection and Classification*. PhD thesis, Old Dominion University.

[Ahmed et al., 2013] Ahmed, H. I., Wei, P., Memon, I., Du, Y., and Xie, W. (2013). Estimation of time difference of arrival (TDOA) for the source radiates bpsk signal. *International Journal of Computer Science Issues (IJCSI)*, 10(3):164.

[Akaike, 1974] Akaike, H. (1974). A new look at the statistical model identification. *IEEE Transactions on Automatic Control*, 19(6):716–723.

[Akula et al., 2016] Akula, R., Kumar, A., and Agrawal, M. (2016). Localization of snow avalanches from infrasound signatures using an acoustic vector sensor. *IUP Journal of Electrical & Electronics Engineering*, 9(2).

[Al-Rashdan and Tahat, 2020] Al-Rashdan, W. Y. and Tahat, A. (2020). A comparative performance evaluation of machine learning algorithms for fingerprinting based localization in dm-mimo wireless systems relying on big data techniques. *IEEE Access*, 8:109522–109534.

[Alaris, 2020] Alaris, A. (2019 (accessed February 3, 2020)). *Antenna DF-A0029 Brochure.*

[Algeier et al., 2010] Algeier, V., Demissie, B., Koch, W., and Thomae, R. (2010). Blind localization of mobile terminals in urban scenarios. ISLE.

[Aljadeff et al., 2010] Aljadeff, D., Amsalem, R., and Amir, L. (2010). Method and system for synchronization offset reduction in a TDOA location system. US Patent 7,729,707.

[Alsheikh et al., 2014] Alsheikh, M. A., Lin, S., Niyato, D., and Tan, H.-P. (2014). Machine learning in wireless sensor networks: Algorithms, strategies, and applications. *IEEE Communications Surveys & Tutorials*, 16(4):1996–2018.

[Apolinário et al., 2019] Apolinário, J. A., Yazdanpanah, H., Nascimento, A., and de Campos, M. L. (2019). A data-selective ls solution to TDOA-based source localization. In *ICASSP 2019-2019 IEEE International Conference on Acoustics, Speech and Signal Processing (ICASSP)*, pages 4400–4404. IEEE.

[Baker and Pratt, 2013] Baker, J. and Pratt, T. G. (2013). A map-based method for geolocation in multipath environments. In *MILCOM 2013-2013 IEEE Military Communications Conference*, pages 58–63. IEEE.

[Balanis, 2012] Balanis, C. A. (2012). *Advanced engineering electromagnetics.* John Wiley & Sons.

[Bedford et al., 2019] Bedford, M. D., Hrovat, A., Kennedy, G., Javornik, T., and Foster, P. (2019). Modeling microwave propagation in natural caves using lidar and ray tracing. *IEEE Transactions on Antennas and Propagation*, 68(5):3878–3888.

[Behrendt et al., 2006] Behrendt, K., Fodero, K., et al. (2006). The perfect time: An examination of time-synchronization techniques. *Publication, Schweitzer Engineering Laboratories, Inc*, pages 1–18.

[Ben-Israel, 1992] Ben-Israel, A. (1992). A volume associated with mxn matrices. *Linear algebra and its applications*, 167:87–111.

[Bennett, 2010] Bennett, J. (2010). *OpenStreetMap.* Packt Publishing Ltd.

[Biau and Scornet, 2016] Biau, G. and Scornet, E. (2016). A random forest guided tour. *Test*, 25(2):197–227.

[Bibb et al., 2016] Bibb, D. A., Yun, Z., and Iskander, M. F. (2016). Machine learning for source localization in urban environments. In *MILCOM 2016-2016 IEEE Military Communications Conference*, pages 401–405. IEEE.

[Bonna, 2017] Bonna, K. Y. (2017). *On the performance of subspace SIMO blind channel identification methods*. PhD thesis, Rutgers University-School of Graduate Studies.

[Botteron et al., 2004] Botteron, C., Host-Madsen, A., and Fattouche, M. (2004). Cramer-Rao bounds for the estimation of multipath parameters and mobiles' positions in asynchronous ds-cdma systems. *IEEE Transactions on Signal Processing*, 52(4):862–875.

[Breiman, 1996] Breiman, L. (1996). Bagging predictors. *Machine learning*, 24(2):123–140.

[Breiman, 2004] Breiman, L. (2004). Consistency for a simple model of random forests.

[Brunato and Battiti, 2005] Brunato, M. and Battiti, R. (2005). Statistical learning theory for location fingerprinting in wireless lans. *Computer Networks*, 47(6):825–845.

[Brutti et al., 2008] Brutti, A., Omologo, M., and Svaizer, P. (2008). Comparison between different sound source localization techniques based on a real data collection. In *2008 hands-free speech communication and microphone arrays*, pages 69–72. IEEE.

[Butt et al., 2020] Butt, M. M., Rao, A., and Yoon, D. (2020). Rf fingerprinting and deep learning assisted ue positioning in 5G. In *2020 IEEE 91st Vehicular Technology Conference (VTC2020-Spring)*, pages 1–7. IEEE.

[Byun et al., 2002] Byun, S. H., Hajj, G. A., and Young, L. E. (2002). Assessment of gps signal multipath interference. In *Proceedings of the 2002 National Technical Meeting of The Institute of Navigation*, pages 694–705.

[Carter, 1993] Carter, C. (1993). Coherence and time delay estimation: An applied tutorial for research, development, test, and evaluation engineers, edited by g. *Clifford Carter (IEEE, Piscataway, NJ, 1993)*, page 506.

[Carter, 1987] Carter, G. C. (1987). Coherence and time delay estimation. *Proceedings of the IEEE*, 75(2):236–255.

[Chan et al., 2008] Chan, F. K., So, H., Zheng, J., and Lui, K. W. (2008). Best linear unbiased estimator approach for time-of-arrival based localisation. *IET Signal Processing*, 2(2):156–162.

[Chan and Ho, 1994] Chan, Y.-T. and Ho, K. (1994). A simple and efficient estimator for hyperbolic location. *IEEE Transactions on Signal Processing*, 42(8):1905–1915.

[Charbonnier et al., 2017] Charbonnier, R., Aslam, M. Z., Corre, Y., and Lostanlen, Y. (2017). Mixing deterministic and stochastic propagation for assessing mmwave small-cell networks. In *2017 11th European Conference on Antennas and Propagation (EUCAP)*, pages 136–140. IEEE.

[Chau, 2002] Chau, C. C. (2002). *Blind Channel Identification for the Emitter Location*

*Problem: A Least Square Approach.* PhD thesis, State University of New York at Binghamton, Electrical Engineering Department.

[Chen, 2012] Chen, C.-S. (2012). Artificial neural network for location estimation in wireless communication systems. *Sensors*, 12(3):2798–2817.

[Cheney and Given, 2017] Cheney, M. and Given, J. A. (2017). Wideband passive source localization. *Inverse Problems*, 33(8):085007.

[Cheng et al., 2012] Cheng, L., Wu, C., Zhang, Y., Wu, H., Li, M., and Maple, C. (2012). A survey of localization in wireless sensor network. *International Journal of Distributed Sensor Networks*, 8(12):962523.

[Cho et al., 2012a] Cho, J. A., Na, H., Kim, S., and Ahn, C. (2012a). Moving-target tracking based on particle filter with TDOA/FDOA measurements. *Etri Journal*, 34(2):260–263.

[Cho et al., 2012b] Cho, S. H., Yeo, S. R., Choi, H. H., Park, C., and Lee, S. J. (2012b). A design of synchronization method for TDOA-based positioning system. In *2012 12th International Conference on Control, Automation and Systems*, pages 1373–1375. IEEE.

[Cline, 1978] Cline, J. (1978). Multilateration error ellipsoids. *IEEE Transactions on Aerospace and Electronic Systems*, (4):665–667.

[Compagnoni et al., 2017a] Compagnoni, M., Canclini, A., Bestagini, P., Antonacci, F., Sarti, A., and Tubaro, S. (2017a). Source localization and denoising: a perspective from the TDOA space. *Multidimensional Systems and Signal Processing*, 28(4):1283–1308.

[Compagnoni et al., 2017b] Compagnoni, M., Pini, A., Canclini, A., Bestagini, P., Antonacci, F., Tubaro, S., and Sarti, A. (2017b). A geometrical–statistical approach to outlier removal for TDOA measurements. *IEEE Transactions on Signal Processing*, 65(15):3960–3975.

[da Costa et al., 2011] da Costa, J. P. C. L., Roemer, F., Haardt, M., and de Sousa, R. T. (2011). Multi-dimensional model order selection. *EURASIP Journal on Advances in Signal Processing*, 2011(1):1–13.

[De Groot et al., 2014] De Groot, E., Bose, T., Cooper, C., and Kruse, M. (2014). Remote transmitter tracking with raytraced fingerprint database. In *2014 IEEE Military Communications Conference*, pages 325–328. IEEE.

[Degen et al., 2014] Degen, C., Govaers, F., and Koch, W. (2014). Emitter localization under multipath propagation using a likelihood function decomposition that is linear in target space. In *17th International Conference on Information Fusion (FUSION)*, pages 1–8. IEEE.

[Demissie, 2011] Demissie, B. (2011). Direct localization and detection of multiple sources in multi-path environments. In *14th International Conference on Information Fusion*, pages 1–8. IEEE.

[Deschamps, 1972] Deschamps, G. A. (1972). Ray techniques in electromagnetics. *Proceedings of the IEEE*, 60(9):1022–1035.

[Dexiu et al., 2018] Dexiu, H., Huang, Z., Zhang, S., and Jianhua, L. (2018). Joint TDOA, FDOA and differential doppler rate estimation: Method and its performance analysis. *Chinese Journal of Aeronautics*, 31(1):137–147.

[Dictionary, 1989] Dictionary, O. E. (1989). Oxford english dictionary. *Simpson, JA & Weiner, ESC*.

[Dietterich, 2000] Dietterich, T. G. (2000). Ensemble methods in machine learning. In *International workshop on multiple classifier systems*, pages 1–15. Springer.

[Dietterich and Kong, 1995] Dietterich, T. G. and Kong, E. B. (1995). Machine learning bias, statistical bias, and statistical variance of decision tree algorithms. Technical report, Citeseer.

[DJI, 2020] DJI, P. (2019 (accessed February 3, 2020)). *DJI Phantom 3 Pro V2.0 Drone professional DJI*.

[El Gemayel et al., 2013] El Gemayel, N., Jakel, H., and Jondral, F. K. (2013). A hybrid TDOA/RSSD geolocation system using the unscented Kalman filter. In *2013 IEEE 78th Vehicular Technology Conference (VTC Fall)*, pages 1–5. IEEE.

[Falk et al., 2003] Falk, J., Handel, P., and Jansson, M. (2003). Effects of frequency and phase errors in electronic warfare TDOA direction-finding systems. In *IEEE Military Communications Conference, 2003. MILCOM 2003.*, volume 1, pages 118–123. IEEE.

[Ferguson et al., 2018] Ferguson, E. L., Williams, S. B., and Jin, C. T. (2018). Sound source localization in a multipath environment using convolutional neural networks. In *2018 IEEE International Conference on Acoustics, Speech and Signal Processing (ICASSP)*, pages 2386–2390. IEEE.

[Fokin et al., 2018] Fokin, G., Kireev, A., and Al-odhari, A. H. A. (2018). TDOA positioning accuracy performance evaluation for arc sensor configuration. In *2018 Systems of Signals Generating and Processing in the Field of on Board Communications*, pages 1–5. IEEE.

[Fowler and Hu, 2008] Fowler, M. L. and Hu, X. (2008). Signal models for TDOA/FDOA estimation. *IEEE Transactions on Aerospace and Electronic Systems*, 44(4):1543–1550.

[Friedlander, 1987] Friedlander, B. (1987). A passive localization algorithm and its accuracy analysis. *IEEE Journal of Oceanic Engineering*, 12(1):234–245.

[Fuchs, 1999] Fuchs, J.-J. (1999). Multipath time-delay detection and estimation. *IEEE Transactions on Signal Processing*, 47(1):237–243.

[Fuschini et al., 2008] Fuschini, F., El-Sallabi, H., Degli-Esposti, V., Vuokko, L., Guiducci, D., and Vainikainen, P. (2008). Analysis of multipath propagation in urban environment through multidimensional measurements and advanced ray tracing simulation. *IEEE Transactions on Antennas and Propagation*, 56(3):848–857.

[Fuschini et al., 2015] Fuschini, F., Vitucci, E. M., Barbiroli, M., Falciasecca, G., and Degli-Esposti, V. (2015). Ray tracing propagation modeling for future small-cell and indoor applications: A review of current techniques. *Radio Science*, 50(6):469–485.

[Gentner et al., 2016] Gentner, C., Ma, B., Ulmschneider, M., Jost, T., and Dammann, A. (2016). Simultaneous localization and mapping in multipath environments. In *2016 IEEE/ION Position, Location and Navigation Symposium (PLANS)*, pages 807–815. IEEE.

[Géron, 2019] Géron, A. (2019). *Hands-on machine learning with Scikit-Learn, Keras, and TensorFlow: Concepts, tools, and techniques to build intelligent systems*. O'Reilly Media.

[Gowdayyanadoddi et al., 2015] Gowdayyanadoddi, N. S., Curran, J. T., Broumandan, A., and Lachapelle, G. (2015). A ray-tracing technique to characterize gps multipath in the frequency domain.

[Gschwendtner et al., 1995] Gschwendtner, B., Wölfle, G., Burk, B., and Landstorfer, F. (1995). Ray tracing vs. ray launching in 3-d microcell modelling.

[Guan et al., 2013] Guan, W., Lu, B., Li, B., and Jiang, P. (2013). A TDOA localization algorithm based on elman neural network for cellular networks. *Information Technology Journal*, 12(23):7143.

[Gui et al., 2010] Gui, G., Wan, Q., Peng, W., and Adachi, F. (2010). Sparse multipath channel estimation using compressive sampling matching pursuit algorithm. *arXiv preprint arXiv:1005.2270*.

[Guo et al., 2018] Guo, X., Ansari, N., Li, L., and Li, H. (2018). Indoor localization by fusing a group of fingerprints based on random forests. *IEEE Internet of Things Journal*, 5(6):4686–4698.

[Haidari et al., 2019] Haidari, S., Moradi, H., Shahabadi, M., and Dehghan, S. M. (2019). A reflection-based rf source localization algorith. *International Journal of Robotics and Automation*, 34(3).

[Haklay and Weber, 2008] Haklay, M. and Weber, P. (2008). Openstreetmap: User-generated street maps. *IEEE Pervasive computing*, 7(4):12–18.

[Hale, 2012] Hale, K. N. (2012). Expanding the use of time/frequency difference of arrival geolocation in the department of defense. Technical report, RAND CORP SANTA MONICA CA.

[He et al., 2017] He, S., Dong, X., and Lu, W.-S. (2017). Localization algorithms for asynchronous time difference of arrival positioning systems. *EURASIP Journal on Wireless Communications and Networking*, 2017(1):1–14.

[Hinteregger et al., 2017] Hinteregger, S., Leitinger, E., and Witrisal, K. (2017). Bayesian multipath channel estimation considering dense multipath. In *Workshop on Dependable Wireless Communications and Localization for the IoT, COST IRACON, Graz, Austria*.

[Hlinka et al., 2014] Hlinka, O., Slučiak, O., Hlawatsch, F., and Rupp, M. (2014). Distributed data fusion using iterative covariance intersection. In *2014 IEEE International Conference on Acoustics, Speech and Signal Processing (ICASSP)*, pages 1861–1865. IEEE.

[Ho, 2012] Ho, K. (2012). Bias reduction for an explicit solution of source localization using TDOA. *IEEE Transactions on Signal Processing*, 60(5):2101–2114.

[Ho and Chan, 1997] Ho, K. and Chan, Y. (1997). Geolocation of a known altitude object from TDOA and FDOA measurements. *IEEE Transactions on Aerospace and Electronic Systems*, 33(3):770–783.

[Ho et al., 2007] Ho, K., Lu, X., and Kovavisaruch, L.-o. (2007). Source localization using TDOA and FDOA measurements in the presence of receiver location errors: Analysis and solution. *IEEE Transactions on Signal Processing*, 55(2):684–696.

[Ho, 1998] Ho, T. K. (1998). The random subspace method for constructing decision forests. *IEEE Transactions on Pattern Analysis and Machine Intelligence*, 20(8):832–844.

[Hoppe et al., 2017] Hoppe, R., Wölfle, G., and Jakobus, U. (2017). Wave propagation and radio network planning software winprop added to the electromagnetic solver package feko. In *2017 International Applied Computational Electromagnetics Society Symposium-Italy (ACES)*, pages 1–2. IEEE.

[Huang and Wu, 2007] Huang, L. and Wu, S. (2007). Low-complexity mdl method for accurate source enumeration. *IEEE signal processing letters*, 14(9):581–584.

[Huang and Benesty, 2007] Huang, Y. A. and Benesty, J. (2007). *Audio signal processing for next-generation multimedia communication systems*. Springer Science & Business Media.

[Hussain, 2017] Hussain, S. (2017). *Efficient ray-tracing algorithms for radio wave propagation in urban environments.* PhD thesis, Dublin City University.

[Ignagni, 2010] Ignagni, M. (2010). Determination of circular and spherical position-error bounds in system performance analysis. *Journal of Guidance, Control, and Dynamics*, 33(4):1301–1305.

[Ilyas and Mahgoub, 2004] Ilyas, M. and Mahgoub, I. (2004). *Handbook of sensor networks: compact wireless and wired sensing systems.* CRC press.

[ITU-R, 2017] ITU-R (2017). Multipath propagation and parameterization of its characteristics (question itu-r 203/3). Recommendation ITU-R P.1407-06, International Telecommunication Union, Geneva.

[ITU-R, 2019] ITU-R (2019). Propagation Data and Prediction Methods for the Planning of Short-range Outdoor Radio-communication Systems and Radio Local Area Networks in the Frequency Range 300 MHz to 100 GHz. Recommendation ITU-R P.1411-10, International Telecommunication Union, Geneva.

[ITU-T, 2000] ITU-T (2000). Scope and Application of Measurement Equipment Specifications Covered in the O-series Recommendations. Recommendation ITU-R P.1411-10, International Telecommunication Union, Geneva.

[IZTGmbh, 2020] IZTGmbh, I. (2019 (accessed February 3, 2020)). *IZT R3000 Technical Description.*

[Jaffe and Wax, 2014] Jaffe, A. and Wax, M. (2014). Single-site localization via maximum discrimination multipath fingerprinting. *IEEE Transactions on Signal Processing*, 62(7):1718–1728.

[Jagoe, 2003] Jagoe, A. (2003). *Mobile location services: The definitive guide.* Prentice Hall Professional.

[Julier and Uhlmann, 2017] Julier, S. and Uhlmann, J. K. (2017). General decentralized data fusion with covariance intersection. In *Handbook of multisensor data fusion*, pages 339–364. CRC Press.

[Kay, 1993] Kay, S. M. (1993). *Fundamentals of statistical signal processing.* Prentice Hall PTR.

[Khalaf-Allah, 2014] Khalaf-Allah, M. (2014). An extended closed-form least-squares solution for three-dimensional hyperbolic geolocation. In *2014 IEEE Symposium on Industrial Electronics & Applications (ISIEA)*, pages 7–11. IEEE.

[Kikuchi et al., 2006] Kikuchi, S., Sano, A., and Tsuji, H. (2006). Blind mobile positioning in urban environment based on ray-tracing analysis. *EURASIP Journal on Advances in Signal Processing*, 2006:1–12.

[Klautau et al., 2018] Klautau, A., Batista, P., González-Prelcic, N., Wang, Y., and Heath, R. W. (2018). 5G mimo data for machine learning: Application to beam-selection using deep learning. In *2018 Information Theory and Applications Workshop (ITA)*, pages 1–9. IEEE.

[Knapp and Carter, 1976] Knapp, C. and Carter, G. (1976). The generalized correlation method for estimation of time delay. *IEEE Transactions on Acoustics, Speech, and Signal Processing*, 24(4):320–327.

[Koch, 2016] Koch, W. (2016). *Tracking and sensor data fusion*. Springer.

[Kolb et al., 2012] Kolb, H., Kleschenkow, A., Warzügel, S., and Kolb, D. (2012). Method and system for synchronization offset reduction in a TDOA location system.

[Koorapaty et al., 1998] Koorapaty, H., Grubeck, H., and Cedervall, M. (1998). Effect of biased measurement errors on accuracy of position location methods. In *IEEE GLOBECOM 1998 (Cat. NO. 98CH36250)*, volume 3, pages 1497–1502. IEEE.

[Kostic et al., 1992] Kostic, Z., Sezan, M. I., and Titlebaum, E. L. (1992). Estimation of the parameters of a multipath channel using set-theoretic deconvolution. *IEEE Transactions on Communications*, 40(6):1006–1011.

[Kovavisaruch and Ho, 2005] Kovavisaruch, L. and Ho, K. (2005). Modified Taylor-series method for source and receiver localization using TDOA measurements with erroneous receiver positions. In *2005 IEEE International Symposium on Circuits and Systems*, pages 2295–2298. IEEE.

[Landmann, 2008] Landmann, M. (2008). *Limitations of experimental channel characterisation*. Grinverl.

[Larew et al., 2013] Larew, S. G., Thomas, T. A., Cudak, M., and Ghosh, A. (2013). Air interface design and ray tracing study for 5G millimeter wave communications. In *2013 IEEE Globecom Workshops (GC WKSHPS)*, pages 117–122. IEEE.

[Lawton and McGeehan, 1994] Lawton, M. C. and McGeehan, J. (1994). The application of a deterministic ray launching algorithm for the prediction of radio channel characteristics in small-cell environments. *IEEE Transactions on Vehicular Technology*, 43(4):955–969.

[Lee et al., 2017] Lee, K., Oh, J., and You, K. (2017). Closed-form solution of TDOA-based geolocation and tracking: A recursive weighted least square approach. *Wireless Personal Communications*, 94(4):3451–3464.

[Leigsnering et al., 2014] Leigsnering, M., Ahmad, F., Amin, M., and Zoubir, A. (2014).

Multipath exploitation in through-the-wall radar imaging using sparse reconstruction. *IEEE Transactions on Aerospace and Electronic Systems*, 50(2):920–939.

[Li, 2006] Li, B. (2006). *Terrestial Mobile User Positioning Using TDOA and Fingerprinting Techniques.* University of New South Wales.

[Li et al., 2017] Li, J., Lu, I.-T., Lu, J. S., and Zhang, L. (2017). Robust kernel-based machine learning localization using nlos TOAs or TDOAs. In *2017 IEEE Long Island Systems, Applications and Technology Conference (LISAT)*, pages 1–6. IEEE.

[Li et al., 2013] Li, Z., Dimitrova, D. C., Braun, T., and Rosário, D. (2013). Highly accurate evaluation of gps synchronization for TDOA localization. In *2013 IFIP Wireless Days (WD)*, pages 1–3. IEEE.

[Li et al., 2014] Li, Z., Dimitrova, D. C., Raluy, D. H., and Braun, T. (2014). TDOA for narrow-band signal with low sampling rate and imperfect synchronization. In *2014 7th IFIP Wireless and Mobile Networking Conference (WMNC)*, pages 1–8. IEEE.

[Liggins II et al., 2017] Liggins II, M., Hall, D., and Llinas, J. (2017). *Handbook of multisensor data fusion: theory and practice.* CRC press.

[Liu et al., 1996] Liu, H., Xu, G., Tong, L., and Kailath, T. (1996). Recent developments in blind channel equalization: From cyclostationarity to subspaces. *Signal Processing*, 50(1-2):83–99.

[Louppe, 2014] Louppe, G. (2014). Understanding random forests: From theory to practice. *arXiv preprint arXiv:1407.7502.*

[Mahfouz et al., 2014] Mahfouz, S., Mourad-Chehade, F., Honeine, P., Farah, J., and Snoussi, H. (2014). Target tracking using machine learning and Kalman filter in wireless sensor networks. *IEEE Sensors Journal*, 14(10):3715–3725.

[Mahler et al., 2015] Mahler, K., Keusgen, W., Tufvesson, F., Zemen, T., and Caire, G. (2015). Propagation of multipath components at an urban intersection. In *2015 IEEE 82nd Vehicular Technology Conference (VTC2015-Fall)*, pages 1–5. IEEE.

[Mani, 2012] Mani, F. (2012). *Improved ray-tracing for advanced radio propagation channel modeling.* PhD thesis, ICTEAM, Université Catholique de Louvain Belgium.

[Martinsson and Johnsson, 2002] Martinsson, D. and Johnsson, T. (2002). A deterministic channel model for simulation of mobile radio communications. Master's thesis, Blekinge Institute of Technology, Department of Telecommunications and Signal Processing.

[Masmoudi et al., 2012] Masmoudi, A., Bellili, F., Affes, S., and Stéphenne, A. (2012). A maximum likelihood time delay estimator in a multipath environment using importance sampling. *IEEE Transactions on Signal Processing*, 61(1):182–193.

[Mayyala et al., 2017] Mayyala, Q., Abed-Meraim, K., and Zerguine, A. (2017). Structure-based subspace method for multichannel blind system identification. *IEEE Signal Processing Letters*, 24(8):1183–1187.

[Meng et al., 2016] Meng, W., Xie, L., and Xiao, W. (2016). Optimal TDOA sensor-pair placement with uncertainty in source location. *IEEE Transactions on Vehicular Technology*, 65(11):9260–9271.

[Miao and Ben-Israel, 1992] Miao, J. and Ben-Israel, A. (1992). On principal angles between subspaces in rn. *Linear algebra and its applications*, 171:81–98.

[Mironov and Voroshilin, 2012] Mironov, M. and Voroshilin, E. (2012). The method to increase the estimation accuracy of time differences of arrival at the cost of propagation channel features. In *Anti-counterfeiting, Security, and Identification*, pages 1–4. IEEE.

[Mironovova and Havlis, 2011] Mironovova, M. and Havlis, H. (2011). Calculation of gdop coefficient. *CTU in Prague: Faculty of Electrical Engineering*.

[Mitchell et al., 2011] Mitchell, L., Sloan, T. M., Mewissen, M., Ghazal, P., Forster, T., Piotrowski, M., and Trew, A. S. (2011). A parallel random forest classifier for r. In *Proceedings of the second international workshop on Emerging computational methods for the life sciences*, pages 1–6.

[Mitchell et al., 1997] Mitchell, T. M. et al. (1997). Machine learning.

[Mittermayer, 2014] Mittermayer, M. (2014). *Detection and tracking of mobile channel impulse responses*. PhD thesis.

[Montminy, 2007] Montminy, M. B. (2007). Passive geolocation of low power emitters in urban environments using TDOA.

[Moulines et al., 1995] Moulines, E., Duhamel, P., Cardoso, J.-F., and Mayrargue, S. (1995). Subspace methods for the blind identification of multiannel fir filters. *IEEE Transactions on signal processing*, 43(2):516–525.

[Musicki et al., 2009] Musicki, D., Kaune, R., and Koch, W. (2009). Mobile emitter geolocation and tracking using TDOA and FDOA measurements. *IEEE Transactions on Signal Processing*, 58(3):1863–1874.

[Naseri and Koivunen, 2016] Naseri, H. and Koivunen, V. (2016). Cooperative simultaneous localization and mapping by exploiting multipath propagation. *IEEE Transactions on Signal Processing*, 65(1):200–211.

[Navabi et al., 2018] Navabi, S., Wang, C., Bursalioglu, O. Y., and Papadopoulos, H. (2018). Predicting wireless channel features using neural networks. In *2018 IEEE*

*International Conference on Communications (ICC)*, pages 1–6. IEEE.

[Neter et al., 1989] Neter, J., Wasserman, W., and Kutner, M. H. (1989). Applied linear regression models.

[Nguyen et al., 2005] Nguyen, X., Jordan, M. I., and Sinopoli, B. (2005). A kernel-based learning approach to ad hoc sensor network localization. *ACM Transactions on Sensor Networks (TOSN)*, 1(1):134–152.

[NI, 2014] NI, U. (2014). 292x/293x Datasheet. *National Instruments.*

[Noack et al., 2017a] Noack, B., Sijs, J., and Hanebeck, U. D. (2017a). Inverse covariance intersection: New insights and properties. In *2017 20th International Conference on Information Fusion (Fusion)*, pages 1–8. IEEE.

[Noack et al., 2017b] Noack, B., Sijs, J., Reinhardt, M., and Hanebeck, U. D. (2017b). Decentralized data fusion with inverse covariance intersection. *Automatica*, 79:35–41.

[Noroozi et al., 2017] Noroozi, A., Oveis, A. H., Hosseini, S. M., and Sebt, M. A. (2017). Improved algebraic solution for source localization from TDOA and FDOA measurements. *IEEE Wireless Communications Letters*, 7(3):352–355.

[O'Donoughue, 2019] O'Donoughue, N. (2019). *Emitter Detection and Geolocation for Electronic Warfare.* Artech House.

[Osugi et al., 2005] Osugi, T., Kim, D., and Scott, S. (2005). Balancing exploration and exploitation: A new algorithm for active machine learning. In *Fifth IEEE International Conference on Data Mining (ICDM'05)*, pages 8–pp. IEEE.

[Pan et al., 2006] Pan, J. J., Kwok, J. T., Yang, Q., and Chen, Y. (2006). Multidimensional vector regression for accurate and low-cost location estimation in pervasive computing. *IEEE Transactions on Knowledge and Data Engineering*, 18(9):1181–1193.

[Park et al., 2017] Park, Y., Seong, W., and Choo, Y. (2017). Compressive time delay estimation off the grid. *The Journal of the Acoustical Society of America*, 141(6):EL585–EL591.

[Phelan et al., 2012] Phelan, B. R., Lenzing, E. H., and Narayanan, R. M. (2012). Source localization using unique characterizations of multipath propagation in an urban environment. In *2012 IEEE 7th Sensor Array and Multichannel Signal Processing Workshop (SAM)*, pages 189–192. IEEE.

[Poisel, 2008] Poisel, R. A. (2008). Introduction to communication electronic warfare systems. artech house. *Inc., Norwood, MA, USA*, 2.

[Porretta et al., 2003] Porretta, M., Nepa, P., Manara, G., Giannetti, F., Dohler, M., Allen, B., and Aghvami, A. H. (2003). User positioning technique for microcellular wireless networks. *Electronics Letters*, 39(9):745–747.

[Porretta et al., 2004] Porretta, M., Nepa, P., Manara, G., Giannetti, F., Dohler, M., Allen, B., and Aghvami, A. H. (2004). A novel single base station location technique for microcellular wireless networks: description and validation by a deterministic propagation model. *IEEE Transactions on Vehicular Technology*, 53(5):1502–1514.

[Pourhomayoun and Fowler, 2014] Pourhomayoun, M. and Fowler, M. L. (2014). Distributed computation for direct position determination emitter location. *IEEE Transactions on Aerospace and Electronic Systems*, 50(4):2878–2889.

[Qu et al., 2017] Qu, X., Xie, L., and Tan, W. (2017). Iterative constrained weighted least squares source localization using TDOA and FDOA measurements. *IEEE Transactions on Signal Processing*, 65(15):3990–4003.

[Quan, 2017] Quan, Y. (2017). *A new machine learning based method for multi-GNSS data quality assurance and multipath detection*. PhD thesis, University of Nottingham.

[Rao et al., 1973] Rao, C. R., Rao, C. R., Statistiker, M., Rao, C. R., and Rao, C. R. (1973). *Linear statistical inference and its applications*, volume 2. Wiley New York.

[Raschka, 2015] Raschka, S. (2015). *Python machine learning*. Packt publishing ltd.

[Reed et al., 1998] Reed, J. H., Krizman, K. J., Woerner, B. D., and Rappaport, T. S. (1998). An overview of the challenges and progress in meeting the e-911 requirement for location service. *IEEE Communications Magazine*, 36(4):30–37.

[Rehman et al., 2014] Rehman, S. U., Sowerby, K. W., Alam, S., and Ardekani, I. (2014). Radio frequency fingerprinting and its challenges. In *2014 IEEE Conference on Communications and Network Security*, pages 496–497. IEEE.

[Reinhardt et al., 2015] Reinhardt, M., Noack, B., Arambel, P. O., and Hanebeck, U. D. (2015). Minimum covariance bounds for the fusion under unknown correlations. *IEEE Signal Processing Letters*, 22(9):1210–1214.

[Reza, 2000] Reza, R. I. (2000). *Data fusion for improved TOA/TDOA position determination in wireless systems*. PhD thesis, Virginia Tech.

[Richter, 2005] Richter, A. (2005). Estimation of radio channel parameters: Models and algorithms. ISLE.

[Richter and Thoma, 2005] Richter, A. and Thoma, R. S. (2005). Joint maximum likelihood estimation of specular paths and distributed diffuse scattering. In *2005 IEEE 61st Vehicular Technology Conference*, volume 1, pages 11–15. IEEE.

[Roivainen et al., 2016] Roivainen, A., Dias, C. F., Tervo, N., Hovinen, V., Sonkki, M., and Latva-aho, M. (2016). Geometry-based stochastic channel model for two-story lobby environment at 10 ghz. *IEEE Transactions on Antennas and Propagation*, 64(9):3990–4003.

[Rottensteiner and Briese, 2002] Rottensteiner, F. and Briese, C. (2002). A new method for building extraction in urban areas from high-resolution lidar data. In *International Archives of Photogrammetry Remote Sensing and Spatial Information Sciences*, volume 34, pages 295–301. Natural Resources Canada.

[Salari et al., 2018] Salari, S., Chan, F., Chan, Y.-T., and Read, W. (2018). TDOA estimation with compressive sensing measurements and hadamard matrix. *IEEE Transactions on Aerospace and Electronic Systems*, 54(6):3137–3142.

[Sato, 1975] Sato, Y. (1975). A method of self-recovering equalization for multilevel amplitude-modulation systems. *IEEE T*, 23(6):679–682.

[Schmitz et al., 2015] Schmitz, J., Schröder, F., and Mathar, R. (2015). TDOA fingerprinting for localization in non-line-of-sight and multipath environments. In *2015 International Symposium on Antennas and Propagation (ISAP)*, pages 1–4. IEEE.

[Schölkopf et al., 2002] Schölkopf, B., Smola, A. J., Bach, F., et al. (2002). *Learning with kernels: support vector machines, regularization, optimization, and beyond*. MIT press.

[Schumann et al., 2017] Schumann, O., Wöhler, C., Hahn, M., and Dickmann, J. (2017). Comparison of random forest and long short-term memory network performances in classification tasks using radar. In *2017 Sensor Data Fusion: Trends, Solutions, Applications (SDF)*, pages 1–6. IEEE.

[Schwarz, 2020] Schwarz, R. (2019 (accessed February 3, 2020)). *R&S SMU200A Vector Signal Generator - Overview Rohde & Schwarz*.

[Series, 2015] Series, P. (2015). Propagation data and prediction methods for the planning of short-range outdoor radiocommunication systems and radio local area networks in the frequency range 300 mhz to 100 ghz. *tech. rep., ITU, Tech. Rep. ITU-R*.

[Serra et al., 2009] Serra, J., Blanco, L., and Nájar, M. (2009). Cramer-Rao bound for time-delay estimation in the frequency domain. In *2009 17th European Signal Processing Conference*, pages 1037–1041. IEEE.

[Setlur et al., 2013a] Setlur, P., Alli, G., and Nuzzo, L. (2013a). Multipath exploitation in through-wall radar imaging via point spread functions. *IEEE Transactions on Image Processing*, 22(12):4571–4586.

[Setlur et al., 2011] Setlur, P., Amin, M., and Ahmad, F. (2011). Multipath model and exploitation in through-the-wall and urban radar sensing. *IEEE Transactions on Geoscience and Remote Sensing*, 49(10):4021–4034.

[Setlur et al., 2013b] Setlur, P., Negishi, T., Devroye, N., and Erricolo, D. (2013b). Multipath exploitation in non-los urban synthetic aperture radar. *IEEE Journal of Selected Topics in Signal Processing*, 8(1):137–152.

[Shi et al., 2016] Shi, H., Zhang, H., and Wang, X. (2016). A TDOA technique with super-resolution based on the volume cross-correlation function. *IEEE Transactions on Signal Processing*, 64(21):5682–5695.

[Shrivastava et al., 2017] Shrivastava, A., Pfister, T., Tuzel, O., Susskind, J., Wang, W., and Webb, R. (2017). Learning from simulated and unsupervised images through adversarial training. In *Proceedings of the IEEE Conference on Computer Vision and Pattern Recognition*, pages 2107–2116.

[Sinha and Hwang, 2020] Sinha, R. S. and Hwang, S.-H. (2020). Improved rssi-based data augmentation technique for fingerprint indoor localisation. *Electronics*, 9(5):851.

[SMT, 2020] SMT, . (2016 (accessed February 3, 2020)). *SAAB Medav Radio Monitoring KCS Product-Sheet.*

[Sohn et al., 2007] Sohn, B., Lee, J., Chae, H., and Yu, W. (2007). Localization system for mobile robot using wireless communication with ir landmark. In *Proceedings of the 1st International Conference on Robot Communication and Coordination*, pages 1–6. Citeseer.

[Stein, 1981] Stein, S. (1981). Algorithms for ambiguity function processing. *IEEE Transactions on Acoustics, Speech, and Signal Processing*, 29(3):588–599.

[Stein, 1993] Stein, S. (1993). Differential delay/doppler ml estimation with unknown signals. *IEEE Transactions on Signal Processing*, 41(8):2717–2719.

[Steinberg and Bowman, 2017] Steinberg, A. N. and Bowman, C. L. (2017). Revisions to the jdl data fusion model. In *Handbook of multisensor data fusion*, pages 65–88. CRC press.

[Steinböck et al., 2016] Steinböck, G., Karstensen, A., Kyösti, P., and Hekkala, A. (2016). A 5G hybrid channel model considering rays and geometric stochastic propagation graph. In *2016 IEEE 27th Annual International Symposium on Personal, Indoor, and Mobile Radio Communications (PIMRC)*, pages 1–6. IEEE.

[Stoica et al., 1989] Stoica, P., Moses, R. L., Friedlander, B., and Soderstrom, T. (1989). Maximum likelihood estimation of the parameters of multiple sinusoids from noisy measurements. *IEEE Transactions on Acoustics, Speech, and Signal Processing*, 37(3):378–392.

[Stoica and Nehorai, 1989] Stoica, P. and Nehorai, A. (1989). Music, maximum likelihood, and Cramer-Rao bound. *IEEE Transactions on Acoustics, Speech, and Signal*

*Processing*, 37(5):720–741.

[Stojmenovic, 2005] Stojmenovic, I. (2005). *Handbook of sensor networks: algorithms and architectures*, volume 49. John Wiley & Sons.

[Suresh et al., 2017] Suresh, K., Silva, S., Votion, J., and Cao, Y. (2017). Multi-sensor data pattern recognition for multi-target localization: A machine learning approach. *arXiv preprint arXiv:1703.00084*.

[Syldatk, 2012] Syldatk, M. (2012). *On Calibration of Ground Sensor Networks*. PhD thesis, Linköping University Electronic Press.

[Tas, 2012] Tas, V. (2012). Optimal use of TDOA geo-location techniques within the mountainous terrain of turkey. Technical report, NAVAL POSTGRADUATE SCHOOL MONTEREY CA.

[Taygur et al., 2018] Taygur, M. M., Sukharevsky, I. O., and Eibert, T. F. (2018). A bidirectional ray-tracing method for antenna coupling evaluation based on the reciprocity theorem. *IEEE Transactions on Antennas and Propagation*, 66(12):6654–6664.

[Taylor, 2013] Taylor, R. C. (2013). *Received signal strength-based localization of non-collaborative emitters in the presence of correlated shadowing*. PhD thesis, Virginia Tech.

[Thoma et al., 2019] Thoma, R. S., Andrich, C., Del Galdo, G., Dobereiner, M., Hein, M. A., Kaske, M., Schafer, G., Schieler, S., Schneider, C., Schwind, A., et al. (2019). Cooperative passive coherent location: A promising 5G service to support road safety. *IEEE Communications Magazine*, 57(9):86–92.

[Torrieri, 1984] Torrieri, D. J. (1984). Statistical theory of passive location systems. *IEEE transactions on Aerospace and Electronic Systems*, (2):183–198.

[Tran and Nguyen, 2008] Tran, D. A. and Nguyen, T. (2008). Localization in wireless sensor networks based on support vector machines. *IEEE Transactions on Parallel and Distributed Systems*, 19(7):981–994.

[Valaee and Kabal, 1995] Valaee, S. and Kabal, P. (1995). Wideband array processing using a two-sided correlation transformation. *IEEE Transactions on Signal processing*, 43(1):160–172.

[Van Nguyen et al., 2015] Van Nguyen, T., Jeong, Y., Shin, H., and Win, M. Z. (2015). Machine learning for wideband localization. *IEEE Journal on Selected Areas in Communications*, 33(7):1357–1380.

[Vaughan and Andersen, 2003] Vaughan, R. and Andersen, J. B. (2003). *Channels, propagation and antennas for mobile communications*. Number 50. Iet.

[Velasco et al., 2016] Velasco, J., Pizarro, D., Macias-Guarasa, J., and Asaei, A. (2016). TDOA matrices: Algebraic properties and their application to robust denoising with missing data. *IEEE Transactions on signal processing*, 64(20):5242–5254.

[Vin et al., 2014] Vin, I., Gaillot, D. P., Laly, P., Lienard, M., and Degauque, P. (2014). Multipath component distance-based fingerprinting technique for non-cooperative outdoor localization in nlos scenarios. *IEEE Transactions on Antennas and Propagation*, 62(9):4794–4798.

[Vin et al., 2015] Vin, I., Gaillot, D. P., Laly, P., Liénard, M., and Degauque, P. (2015). Overview of mobile localization techniques and performances of a novel fingerprinting-based method. *Comptes Rendus Physique*, 16(9):862–873.

[Vin, 2014] Vin, I. K. (2014). *Algorithmes de localisation de mobiles en milieu urbain en mode non-coopératif.* PhD thesis, Lille 1.

[Walck et al., 2007] Walck, C. et al. (2007). Hand-book on statistical distributions for experimentalists. *University of Stockholm*, 10.

[Wang et al., 2016] Wang, W., Wang, G., Zhang, F., and Li, Y. (2016). Second-order cone relaxation for TDOA-based localization under mixed los/nlos conditions. *IEEE Signal Processing Letters*, 23(12):1872–1876.

[Wang et al., 2018] Wang, Y., Xiu, C., Zhang, X., and Yang, D. (2018). Wifi indoor localization with csi fingerprinting-based random forest. *Sensors*, 18(9):2869.

[Weiss and Stein, 1987] Weiss, A. J. and Stein, Z. (1987). Optimal below threshold delay estimation for radio signals. *IEEE transactions on aerospace and electronic systems*, (6):726–730.

[Whitehouse, 2002] Whitehouse, K. (2002). *The design of calamari: an ad-hoc localization system for sensor networks.* PhD thesis, Masterâ€™s thesis, University of California at Berkeley.

[Wielandt and Strycker, 2017] Wielandt, S. and Strycker, L. D. (2017). Indoor multipath assisted angle of arrival localization. *Sensors*, 17(11):2522.

[Wong et al., 2016] Wong, S. C., Gatt, A., Stamatescu, V., and McDonnell, M. D. (2016). Understanding data augmentation for classification: when to warp? In *2016 International Conference on Digital Image Computing: Techniques and Applications (DICTA)*, pages 1–6. IEEE.

[Wood et al., 2016] Wood, E., Baltrušaitis, T., Morency, L.-P., Robinson, P., and Bulling, A. (2016). Learning an appearance-based gaze estimator from one million synthesised images. In *Proceedings of the Ninth Biennial ACM Symposium on Eye Tracking Research & Applications*, pages 131–138.

[Wymeersch et al., 2009] Wymeersch, H., Lien, J., and Win, M. Z. (2009). Cooperative localization in wireless networks. *Proceedings of the IEEE*, 97(2):427–450.

[Xie et al., 2018] Xie, J., Song, Z., Li, Y., Zhang, Y., Yu, H., Zhan, J., Ma, Z., Qiao, Y., Zhang, J., and Guo, J. (2018). A survey on machine learning-based mobile big data analysis: Challenges and applications. *Wireless Communications and Mobile Computing*, 2018.

[Xu et al., 2012] Xu, B., Sun, G., Yu, R., and Yang, Z. (2012). High-accuracy TDOA-based localization without time synchronization. *IEEE Transactions on Parallel and Distributed Systems*, 24(8):1567–1576.

[Yin, 2014] Yin, F. (2014). Robust wireless localization in harsh mixed line-of-sight/non-line-of-sight environments.

[Yu et al., 2017] Yu, X., Wu, X., Luo, C., and Ren, P. (2017). Deep learning in remote sensing scene classification: a data augmentation enhanced convolutional neural network framework. *GIScience & Remote Sensing*, 54(5):741–758.

[Yun and Iskander, 2015] Yun, Z. and Iskander, M. F. (2015). Ray tracing for radio propagation modeling: Principles and applications. *IEEE Access*, 3:1089–1100.

[Zafari et al., 2019] Zafari, F., Gkelias, A., and Leung, K. K. (2019). A survey of indoor localization systems and technologies. *IEEE Communications Surveys & Tutorials*, 21(3):2568–2599.

[Zekavat and Buehrer, 2011] Zekavat, R. and Buehrer, R. M. (2011). *Handbook of position location: Theory, practice and advances*, volume 27. John Wiley & Sons.

[Zeng et al., 2010] Zeng, W.-J., Zhang, X.-D., Li, X.-L., and Cheng, E. (2010). Multipath time-of-arrival estimation via modified projection onto convex sets. In *2010 IEEE International Conference on Acoustics, Speech and Signal Processing*, pages 3226–3229. IEEE.

[Zhang et al., 2016] Zhang, X., Tepedelenlioglu, C., Banavar, M., and Spanias, A. (2016). Node localization in wireless sensor networks. *Synthesis Lectures on Communications*, 9(1):1–62.

[Zhao et al., 2014] Zhao, Y., Liu, K., Ma, Y., and Li, Z. (2014). An improved k-nn algorithm for localization in multipath environments. *EURASIP Journal on Wireless Communications and Networking*, 2014(1):1–10.

[Zhu et al., 2018] Zhu, Y., Deng, B., Jiang, A., Liu, X., Tang, Y., and Yao, X. (2018). Admm-based TDOA estimation. *IEEE Communications Letters*, 22(7):1406–1409.

[Ziedan, 2012] Ziedan, N. I. (2012). Pattern recognition-based environment identification for robust wireless devices positioning. *Pattern Recognition*, 3(12).

[Zou et al., 2017] Zou, Y., Liu, H., and Wan, Q. (2017). An iterative method for moving target localization using TDOA and FDOA measurements. *IEEE Access*, 6:2746–2754.

# Abbreviations, Symbols and Notation

## Abbreviations

**AoA** Angle Of Arrival

**ANN** Artificial Neural Network

**BSS** Blind Source Separation

**BW** Bandwidth

**CAF** Complex Ambiguity Function

**CBB** Complex Baseband

**CEP** Circular Error Probability

**CDF** Cumulative Distribution Function

**CIR** Channel Impulse Response

**CPCL** Cooperative Passive Coherent Location

**CPSD** Cross Power Spectral Density

**CEP** Circular Error Probability

**CRLB** Cramér-Rao Lower Bound

**DMC** Dense Multi-path Component

**DF** Direction Finding

**DoA** Direction Of Arrival

**DoD** Direction Of Departure

**EIRP** Effective Isotropic Radiated Power

**FDoA** Frequency-Difference Of Arrival

**FIM** Fisher Information Matrix

**FIR** Finite Impulsive Response

**GDOP** Geometric Dilution of Precision

**GIS** Geographic Information System

**GO** Geometrical Optics

**GDOP** Geometric Dilution of Precision

**GPS** Global Positioning System

**GTD** Geometrical Theory of Diffraction

**IoT** Internet of the Things

**ICI** Inverse Covariance Intersection

**LOS** Line Of Sight

**LS** Least Squares

**LIDAR** Light Detection And Ranging

**LLS** Linear Least Square

**LOS** Line Of Sight

**LS** Least Square

**ML** Maximum Likelihood

**MoM** Method of Moments

**MUSIC** Multiple Signal Classification

**ML** Maximum Likehood

**NLOS** Non-Line Of Sight

**NLS** Non-Linear Leas Squared

**PDF** Probability Density Function

**PO** Physical Optics

**PRBS** pseudorandom binary sequence

**RF** Radio Frequency

**RMS** Root Mean Square

**RMSE** Root Mean Squared Error

**RSS** Received Signal Strength

**RX** Receiver

**RT** Ray Tracing

**RL** Ray Launching

**SBR** Shooting and Bouncing Ray

**SC** Specular Component

**SLAM** Simultaneous Localization and Mapping

**SIMO** Single Input Multiple Output

**SNR** Signal to Noise Ratio

**SVD** Singular Value Decomposition

**SVM** Support Vector Machine

**ToA** Time Of Arrival

**TDoA** Time-Difference Of Arrival

**TUI** Technische Universität Ilmenau

**TX** Transmitter

**UAV** Unmanned Aerial Vehicle

**UTD** Uniform Theory of Diffraction

**V2X** Vehicle-to-Vehicle/Infrastructure

**VCC** Volume Cross-Correlation

**WLS** Weighted Least Square

**WLLS** Weighted Linear Least Square

**WSN** Wireless Sensor Network

# Symbols and Notation

$\mathrm{mat}\,A$ Matrix $\mathrm{mat}\,A$

$\boldsymbol{a}$ Vector $\boldsymbol{a}$

$\mathrm{mat}\,A^{\mathrm{H}}$ Hermitian operation of matrix $\mathrm{mat}\,A$

$\mathrm{mat}\,A^{\mathrm{T}}$ Transpose operation of matrix $\mathrm{mat}\,A$

$\det\{\mathrm{mat}\,A\}$ Determinant of the square matrix $\mathrm{mat}\,A$

$|\,\mathrm{mat}\,A\,|$ Absolute values of matrix $\mathrm{mat}\,A$

$\angle\mathrm{mat}\,Q$ Phases of matrix $\mathrm{mat}\,A$

$\mathrm{tr}\{\mathrm{mat}\,A\}$ Trace of the matrix $\mathrm{mat}\,A$

$\mathrm{mat}\,A^{-1}$ Inverse of the matrix $\mathrm{mat}\,A$

$\mathrm{mat}\,I$ Identity matrix

$\log$ base-10 logarithm

$\lambda$ Free-space wavelength

$\sum$ Sum

$\epsilon_0 = 8.8542 \times 10^{-12}\mathrm{F/m}$ Free-space permittivity

$\mu_0 = 4\pi \times 10^{-7}\mathrm{N/A}^2$ Free-space permeability

$\eta_0 = \sqrt{\dfrac{\mu_0}{\epsilon_0}} = 120\pi\,\Omega$ Intrinsic impedance

$k_0 = \dfrac{2\pi}{\lambda} = \omega\sqrt{\mu_0\epsilon_0}$ Free-space wavenumber

$\vartheta$ Elevation angle in degrees

$\varphi$ Azimuth angle in degrees

$E\{x\}$ Expected value of the random variable $x$